Artificial Intelligence Systems in Environmental Engineering

Specialists are paying attention to the growing concern for environmental issues. ***Artificial Intelligence Systems in Environmental Engineering*** uses techniques from the field of Artificial Intelligence as the alternative to address problems that are difficult to model by analytical methods. The book provides a general study and introduces a set of techniques for environmental engineering. With a particular focus on climate change and energy policy, it discusses innovative solutions and models that can positively influence and increase the efficiency of resource use and decrease the impact on the environment while developing the well-being of individuals. The book will be of most interest to research students who have a scientific approach and deal with problems related to environmental engineering.

Jamal Mabrouki, PhD, Assistant Professor at University Mohammed V in Rabat, Morocco.

Azrour Maroude, PhD, Assistant Professor of Computer Science, Moulay Ismail University, Morocco.

Azeem Irshad, PhD, Assistant Professor-Computer Science, Govt. Asghar Mall College Rawalpindi, Pakistan.

Artificial Intelligence Systems in Environmental Engineering

Edited by
Jamal Mabrouki
Azrour Maroude
Azeem Irshad

CRC Press is an imprint of the
Taylor & Francis Group, an **informa** business

Designed cover image: Shutterstock Images

First edition published 2024
by CRC Press
2385 NW Executive Center Drive, Suite 320, Boca Raton FL 33431

and by CRC Press
4 Park Square, Milton Park, Abingdon, Oxon, OX14 4RN

CRC Press is an imprint of Taylor & Francis Group, LLC

ISBN: 978-1-032-56585-9 (hbk)
ISBN: 978-1-032-56586-6 (pbk)
ISBN: 978-1-003-43621-8 (ebk)

DOI: 10.1201/9781003436218

Typeset in Minion
by SPi Technologies India Pvt Ltd (Straive)

Contents

Contributors

Halima Ameziane
Ibn Tofail University
Kenitra, Morocco
and
Mohammed V University
Sale, Morocco

Rachid Amiha
Ibn Zohr University
Agadir, Morocco

Mohammed Benchrifa
University Mohammed V in Rabat
Morocco
and
Ibn Tofaïl University—Kenitra-University Campus
Kenitra, Morocco

Najlaa Ben-Lhachemi
University Mohammed V in Rabat
Morocco

Fatima Benradi
Mohammed V University
Sale, Morocco

Youssef Bouchriti
Ibn Zohr University
Agadir, Morocco

Tarik Bouramtane
University Mohammed V in Rabat
Morocco

Fatine Drhimer
University Mohammed V in Rabat
Morocco

Ahmed Elshaikh
University of Khartoum
Khartoum, Sudan

Karima El-Mouhdi
Higher Institute of Nursing Professions and Healthcare Technics
Meknes, Morocco

Khadija El-Moustaqim
Ibn Tofaïl University—Kenitra-University Campus
Kenitra, Morocco

Mohamed Ait Haddou
Ibn Zohr University
Agadir, Morocco

Driss Hmouni
Ibn Tofaïl University—Kenitra-University Campus
Kenitra, Morocco

Belkacem Kabbachi
Ibn Zohr University
Agadir, Morocco

Ilias Kacimi
University Mohammed V in Rabat
Morocco

Nadia Kassou
University Mohammed V in Rabat
Morocco

Jamal Mabrouki
University Mohammed V in Rabat
Morocco

Qisse Najat
University Mohammed V in Rabat
Morocco

Souad Nasrdine
University Mohammed V in Rabat
Morocco

Nordine Nouayti
Abdelmalek Essaadi University
Al-Hoceima, Morocco

Abderrahman Nounah
Mohammed V University
Sale, Morocco

Mohamed Saadi
University Mohammed V in Rabat
Morocco

Miloudia Slaoui
University Mohammed V in Rabat
Morocco

Rachid Tadili
University Mohammed V in Rabat
Morocco

Latifa Taoufiq
University Mohammed V in Rabat
Morocco

CHAPTER 1

A Scanning Electron Microscope (SEM) Observation of Quartz Grains from Issen Wadi, Western High Atlas, Morocco

Implications for the Sedimentary Environment

Mohamed Ait Haddou, Belkacem Kabbachi, Youssef Bouchriti, and Rachid Amiha

Ibn Zohr University, Agadir, Morocco

1.1 INTRODUCTION

Soil erosion is one of the most severe threats to the environment around the world (Mondal et al., 2015; Zhiying and Fang, 2016). It is a worrying environmental challenge in some regions such as Morocco (Ed-Dakiri et al., 2022). The dynamics of a catchment depend essentially on environmental conditions, climate, geology, land use, and control variables related to liquid and solid flows (Rollet, 2007). At a global level, more than 50% of sediment flows in regulated basins are trapped by dams (Vörösmarty et al., 2003). Other studies have also shown that the sedimentological characterization of solid inputs (sands, pebbles, and clays) from rivers gives very good information on eroded areas and sediment origins such as lithological units, soil types, and mineral assemblages (Elmouden et al., 2005; Omdi et al., 2018). The microtextures on the surface of quartz grains provide an insight into the sedimentary history of sediments, and, in particular, specify the nature of the transport agents causing their shaping (Soyer, 1969; Vos et al., 2014). Therefore, about 250 characters have been listed (Mahaney, 2002). The granulometric and petrographic results, completed or confronted with others of an exoscopic nature, will contribute to the mapping of the zones, potentially producing fine sediments responsible for the silting of the Abdelmomen dam. The Issen watershed is located in a region with an arid to semi-arid

DOI: 10.1201/9781003436218-1

climate; it is characterized by its large area, varied terrain relief and climatic processes, its variable topographical and geomorphological factors and different types of lithology. The exoscopic analysis of quartz grains from the Issen basin has contributed to the knowledge of the basin environment and the dynamics of the wadi. They could also be used in comparisons with the other watersheds of the High Atlas and the right bank of the Oued Souss. The objectives of this study include: (1) to give an overview of calcareous rate distribution in detrital sand-grained sediment supply from the Issen basin; (2) to imply the knowledge for interpretation of different shapes and traces affecting in terms of the quartz particles evolution, mainly in the main watercourse; (3) to deduce spatial transport processes of the main quartz particles, relating to erosional factors and currents in the main river and major tributaries. Understanding the structural and lithological setting, terrain characteristics, geomorphological processes, hydrological condition, land use/land cover and vegetation status, and other topographic factors, elevation, slope, and aspect, is essential for understanding degradation mechanisms and designing effective erosion mitigation measures.

1.2 STUDY AREA, DATA AND METHODS

1.2.1 Case Study: Great Watershed of the Issen Wadi

The Issen basin is an important watershed in the Souss catchment, and a tributary of the Souss Wadi in the central western part of Morocco (Ait Haddou et al., 2023). The study area, the Issen catchment, is a dry zone of the downstream section of the Souss that covers about 1300 ha and extends over 60 km in length and 35 km in width (Ait Haddou et al., 2022a). The proposed study basin has a perimeter of 218.17 km, an drainage density of 0.85km^{-1} and a 63 km length of the main river through it. Between the following latitudes and longitudes, in width and with a NE-SW direction, respectively, are 30°64 and 31°07 North and 8.72° and 9.31° West (Figure 1.1). It is the largest sub-basin of the southern

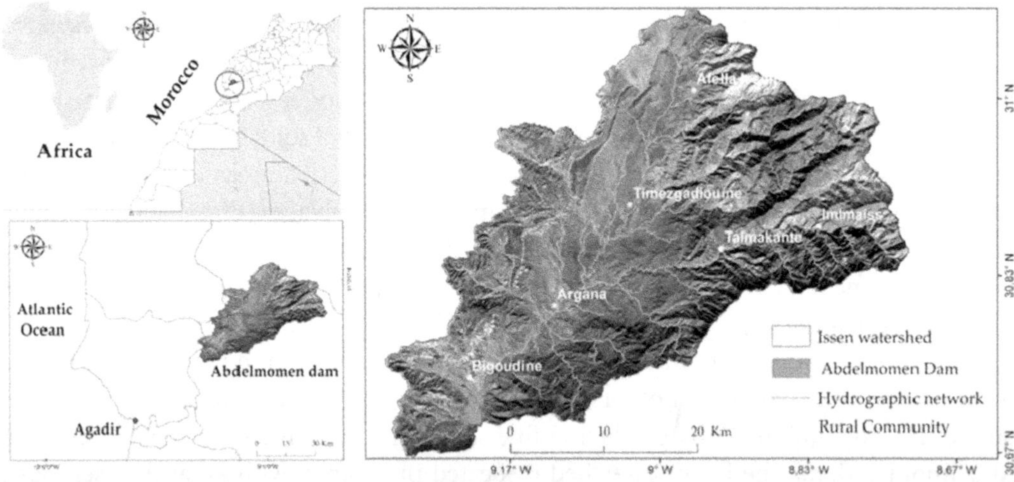

FIGURE 1.1 Location of the Issen Wadi in west central Morocco and its landscape of the Issen river drainage basin.

edge of the Western High Atlas (WHA), which drains the Southern Atlas furrow. At the outlet of this oued, Abdelmomen dam was installed to ensure drinking water supply to about one million persons of the Agadir metropolis, which presents several environmental challenges. Geologically, the Issen basin can be divided into three different zones. It is bounded to the east by the Palaeozoic massif or ancient block of the High Atlas and to the west by the Jurassic and Cretaceous plateaus (Figure 1.2). The tectonics, lithology, and external geodynamics characteristics of the Issen catchment indicate a strongly contrasted morphology (Ait Haddou et al., 2022b).

The climate of the study area is warm and dry, with the mean annual temperature being approximately 21.85°C, The Issen watershed receives an average annual rainfall of 302 mm in the central parts, and the annual evaporation is 2278 mm/year. The average rainfall erosivity is about 537.64 MJ.mm/ha.h.yr for the whole basin (Ait Haddou et al., 2020, Ait Haddou et al., 2022b) (Figure 1.3). The monthly and seasonal distribution of the flows is essentially conditioned by the winter rainfall contributions that are dependent

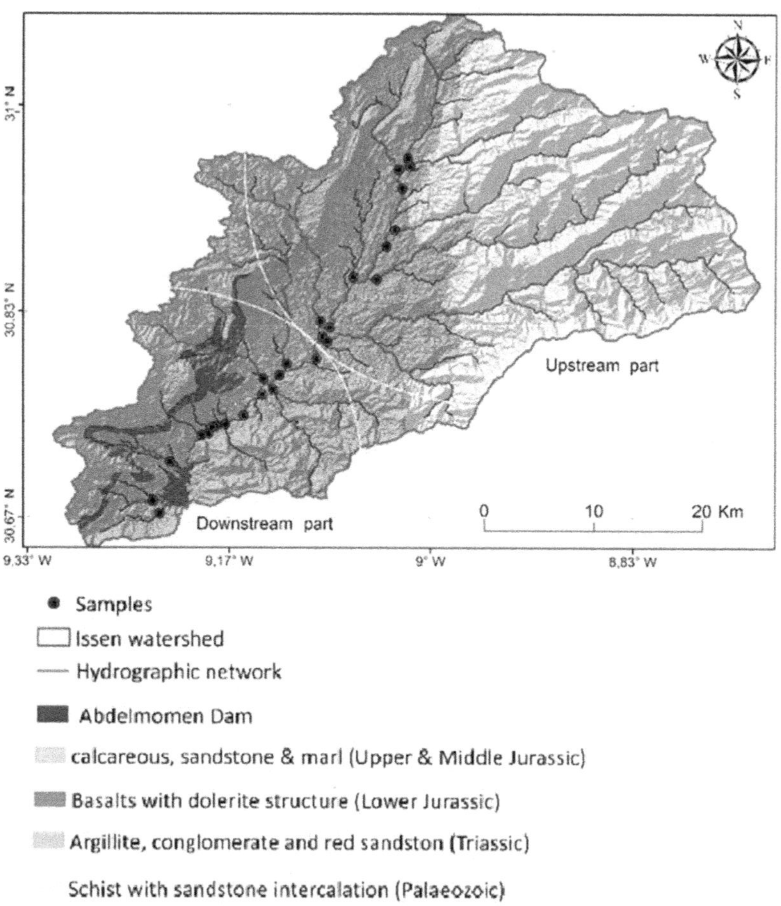

FIGURE 1.2 Simplified geological map and samples localization of Issen basin, extracted from Geological map of Argana corridor (1/100,000) (Tixeront, 1974).

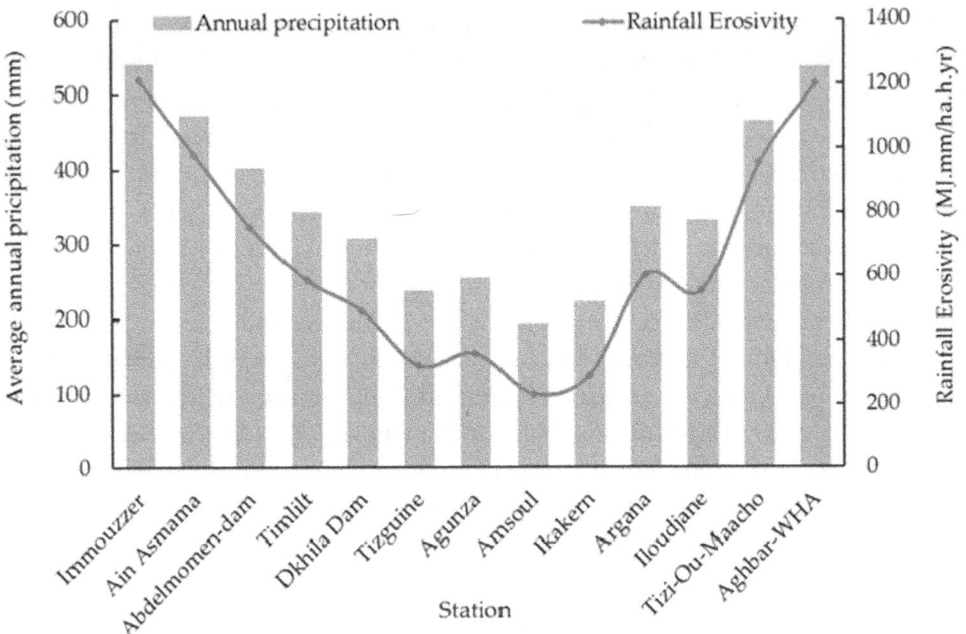

FIGURE 1.3 Distribution of the average annual precipitation and rainfall erosivity in the Issen basin and the adjacent zones. (data sources: ABH, DREFLCD; Ait Haddou et al., 2020, 2022b.)

on the Mediterranean climatic conditions, and by their physical stationary properties of a torrential nature. In addition, the correlation between the calculated interannual means of precipitation of streamflow the 1981–2013 is approximately 70%. The tributaries of the Issen wadi are ephemeral, with an average of several flows per year, mainly in the winter and Automn seasons (September–February). Other results emphasize that the groundwater resources are very limited in the Argana basin (Ait Haddou et al., 2023). Commonly, the lithological facies heterogeneity, soil types, topography, climate, and fluvial system are the most common parameters impacting the spatial hydrological flows' variability.

1.2.2 Deposits Sampling

In this study, the main channel of the river Issen was sampled from the confluence of the rivers Ait Bkhayr and Ait Moussi (Issen) to the Abdelmomen dam. Samples were taken over a distance of 3 to 6 km, depending on the location and physical constraints. The samples were taken by means of a shovel at a depth of 30 cm along the axis of the canal. A quantity of approximately 0.5 kg of sediment was taken and placed in a plastic bag, which was then labeled and reserved for granulometric and exoscopic analyses.

Table 1.1 shows the location of the samples, their GPS coordinates and station characteristics, which are also shown on the map in Figure 1.2.

TABLE 1.1 Samples of the Main Course of the Wadi Issen and the Distance Followed from the Upstream of Issen Basin to the Downstream Side

Samples	Distance in km	Part of the Basin
E24	12.64	Upstream part
E22	16.71	
E20	22.56	
E17	27.64	
E14	33.10	Midstream part
E12	38.04	
E25	41.65	
E9	43.73	Downstream part
E7	48.37	
E5	54	

1.3 METHODS

The determination of the texture allows the observed material to be given the name of a textural class (example: clayey sand, sandy clay, calcareous mud…). These textural classes are defined and represented on "texture triangles". As such, the diagram used follows the example of textural description given in Shepard (1954), while the terminology is taken from Flemming (2015). The granulometric classifications that have been retained are from Blott and Pye (2001). Furthermore, Table 1.2 defines all sand classes according to their carbonate content. Scanning electron microscope (SEM) analysis is a reference method of morphological and particle size distribution (Lee et al., 2019; Bouchriti et al., 2022). They make it possible to determine the depositional environment of grain, its history, and, in some cases, its geographical origin (Fournier et al., 2012). The method used is based on exoscoping quartz grains using a scanning electron microscope, they were used to determine the morphology (shape, sphericity and usur). Sample preparation and SEM imaging are the first, and therefore the most important, steps in the study of surface textural observation (Vos et al., 2014). In order to carry out this study, four samples of sand were taken along the main channel of the Issen, which had already been treated during the granulometric study, the majority of grain coatings and adherent particles are removed by a sample preparation procedure using a bath of 15% diluted hydrochloric acid (HCl) and 50 g/l tetrasodium pyrophosphate solutions. They were then rinsed with distilled water and dried. The scanning electron microscope used in this study is equipped with SEM-EDS JEOL/EO

TABLE 1.2 Classes of Sands According to Their Limestone Percent (%)

Classe	Limestone Percent (%)
Non-calcareous sand	<1
Low calcareous sand	1–5
Moderately calcareous sand	5–15
Highly calcareous sand	15–30
Very highly calcareous	30–50
Extremely calcareous	>50

manufactured by the Japan Electron Optics Laboratory (JEoL). The surface analysis was carried out at the Research Center of the Faculty of Science at Ibn Zohr University, Agadir, Morocco. In addition to the surface state, we have calculated another parameter describing the shape. This is the sphericity of the quartz particles, which allowed us to interpret the evolution of the roundness of the particles through the analysis of SEM images.

1.4 RESULTS AND DISCUSSION

1.4.1 Lithologic Classification

The use of the triangular classification to classify sediment is based on the relative quantities of three mineralogical constituents, namely quartz, clay, and calcite. According to the classification of Table 1.2, all samples along the Issen wadi indicated a low calcareous sand. They ranged from 0.36% to 16.51%, indicating that the carbonate concentration levels in all investigated parts of Issen Wadi were lower than twenty (<20%), indicative of light carbonation, with respect to total of the fractions studied. The variations of limestone frequencies in the three parts of Issen basin are summarized in Table 1.3.

The unconsolidated samples of Wadi Issen are generally not so calcareous, with 65% of the samples being non-calcareous to weakly calcareous. 31% of the samples are moderately

TABLE 1.3 Total Sand, Clay and Limestone Content in the Wadi Issen Samples and Carbonate Levels

Samples	Sand (%)	Clay (%)	Limestone (%)	Interpretation
E24	96.20	0.66	2.96	Low calcareous sand
E23	96.15	0.56	3.02	Low calcareous sand
E21	93.69	0.54	5.63	Moderately calcareous sand
E22	97.54	0.45	1.44	Low calcareous sand
E19	98.78	0.6	0.36	Non-calcareous sand
E20	97.04	1.82	—	Non-calcareous sand
E18	91.31	0.54	7.62	Moderately calcareous sand
E16	97.62	0.51	1.74	Low calcareous sand
E17	98.29	0.28	1.37	Low calcareous sand
E13	98.88	0.15	0.92	Non-calcareous sand
E14	93.65	1.98	3.47	Low calcareous sand
E6	96.22	0.25	3.38	Low calcareous sand
E12	93.30	0.94	4.51	Low calcareous sand
E15	91.91	1.16	6.58	Moderately calcareous sand
E26	86.55	0.77	11.90	Moderately calcareous sand
E11	92.43	0.46	6.45	Moderately calcareous sand
E25	95.30	0.07	4.47	Low calcareous sand
E10	96.29	0.63	2.39	Low calcareous sand
E9	97.58	0.19	2.19	Low calcareous sand
E8	95.76	1.8	1.49	Low calcareous sand
E7	93.43	0.22	6.24	Moderately calcareous sand
E5	93.20	2.12	3.15	Low calcareous sand
E4	83.89	0.89	14.75	Moderately calcareous sand
E3	83.21	0.25	16.51	Highly calcareous sand
E2	89.20	0.78	9.51	Moderately calcareous sand
E1	94.77	0.55	4.52	Low calcareous sand

calcareous and only one sample is strongly calcareous, representing only 4% of the sample. The latter are mainly transported by the tributaries of the right bank of the oued Issen which originate in the Jurassic northern and southern plateaus with essentially carbonate marine and lacustrine sedimentation. This is the case with the Aoujgal tributary E26 in the upstream section of Issen crossing the Ida-Ou-Bouzia cuestas to the north and the three tributaries of the downstream section draining the Ida-Ou-Tanan cuestas to the south and which are E4, E2 and E3, respectively, from Agadir Imouzgaou, Tasadamt and Alemzi. This stream transported strongly carbonated sediments with giant limestone blocks and directly fed the Abdelmomen dam with calcarenites.

If run-off water is responsible for the dissolution of the carbonate formations, the heat will precipitate the limestone contained in the water, thus increasing the granular limestone content of the sands. This is how calcarenites with a very low mud content (1%) are formed. From the above, it can be seen that limestone is present in many samples, but mainly in contents of less than 10%. One could think that two main types of sand exist, those of the left bank poor in limestone and are siliceous sands and those of the right bank richer in limestone and are calcareous sands or calcarenites.

1.4.2 Surface Texture of Quartz Grains

The various sandy materials were photographed using a scanning electron microscope (Figures 1.4–1.7). This allowed the surface of the grains to be qualitatively characterized, in particular their roughness, angularity, and sphericity. The analysis of images taken by SEM or exoscopy allows us to observe and interpret the surface condition of the quartz grains of the four samples analyzed. Thus, all the grains are marked by numerous traces that have distinctive shapes and sizes of the factors responsible for their appearance. The mechanical impacts observed better affect the quartz grains of the downstream section of Issen, they are related to an energetic hydrodynamic context. Dissolution pits can be observed on flat surfaces and in cavities. Strong pitting affects the entire surface of the grain, this indicates long water transport. Thus, the exoscopic analysis revealed analogies in the morphology of the quartz grains. In summary, the surface of the quartz grains of the four samples analyzed is marked by numerous traces that have shapes and sizes characteristic of the factors that formed them. These factors can be of physical, chemical, mechanical, or biological origin.

The total absence of worn forms whose edges show no trace of polishing or rounding does not necessarily indicate that the deposit is recent. However, it does at least allows us not to misinterpret all the grains, especially those where the deposits are fed by the tributaries from upstream Issen to the upper basin (case of some grains of E24 and E17). On the other hand, the commonly impacted polyhedra do not indicate wear by the long path travelled but are rather linked to the impact of the quarries installed along the Oued Issen, as in the case of sample E17. Some grains of samples E17 and E5 show inclusions or impurities which may prove the hypothesis of a replacement following acid dissolution. The crystallization voids can weaken the crystalline integrity of the quartz, thus favouring the location of a mechanical impact which will provide numerous microfractures, fragments and flakes. This further demonstrates the impact of other crystallization and recrystallization factors in the characterization of grain morphology.

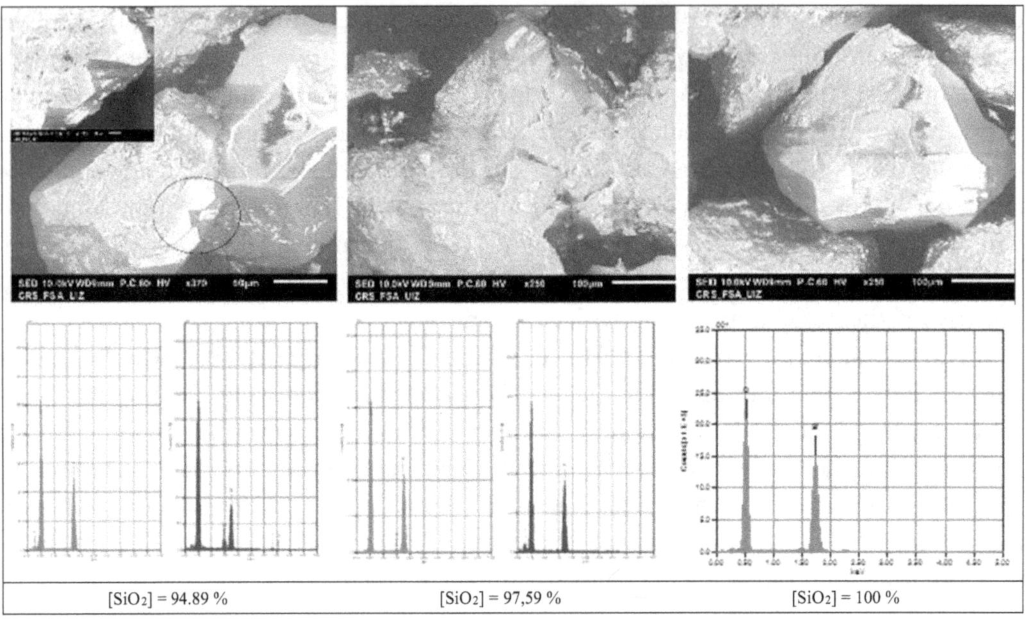

FIGURE 1.4 Examples of SEM photographs of quartz grains showing grain surface characteristics of E24 sample (Issen upstream) with fractures at the level of a fracture whose edge is fresh.

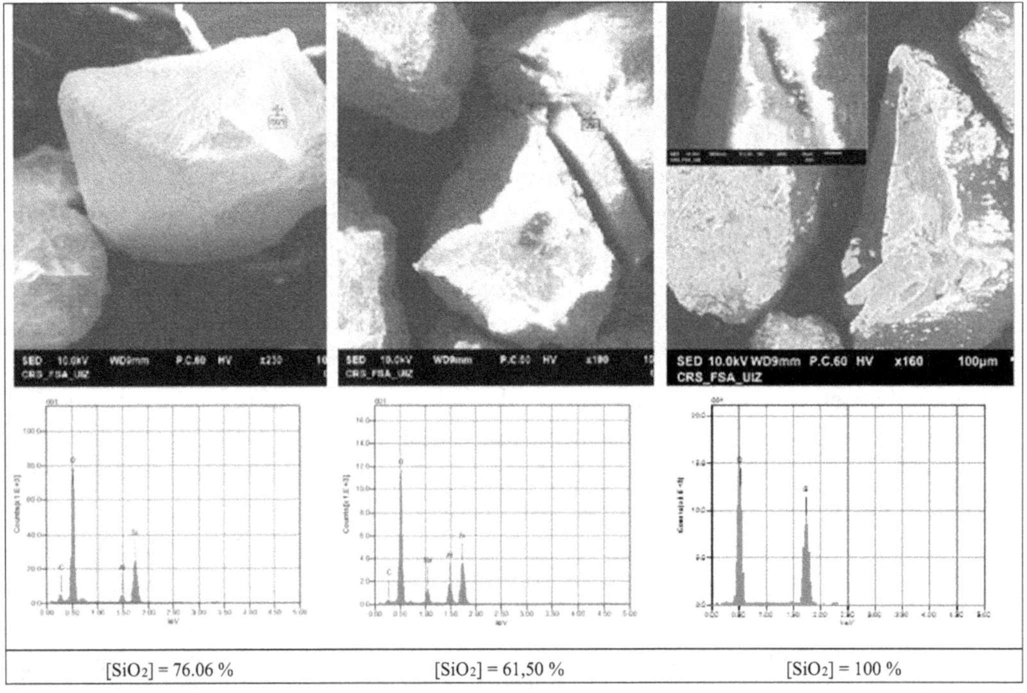

FIGURE 1.5 Exoscopy of the quartz grains of the E17 sample (upstream part). Note the large microfracturation linked to the mechanical action, slightly abraded edges with fresh impact marks.

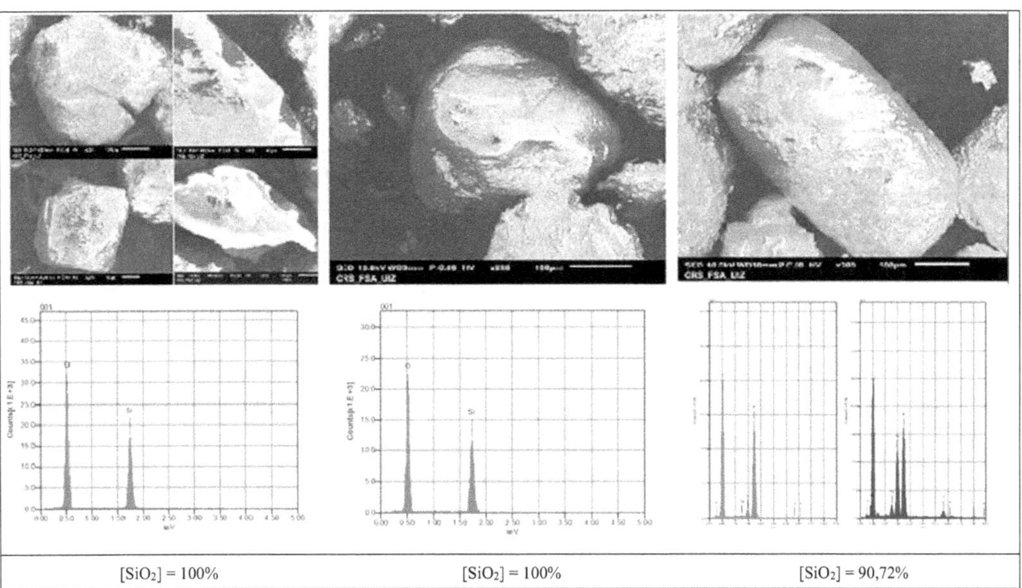

FIGURE 1.6 Exoscopy of the quartz grains of the E12 sample (midstream part of Issen Wadi), showing the polyders and a zone of surface destructuration linked to a mechanical impact.

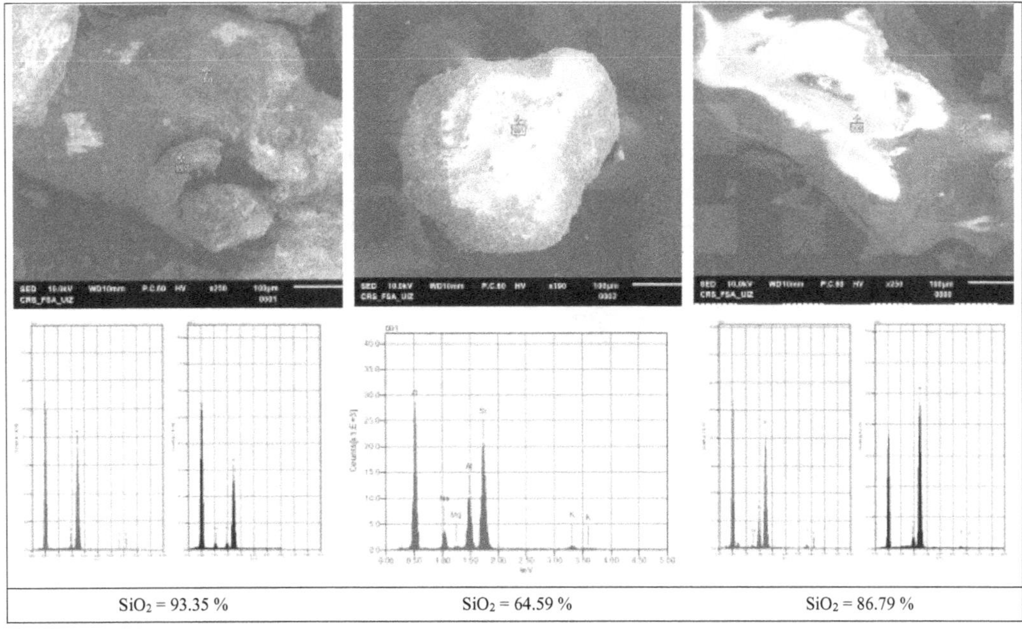

FIGURE 1.7 Quartz grains in the SEM. Note the numerous mechanical impacts following the reworking phenomenon over a long distance (downstream part) (E5).

1.4.3 Quartz Grain Sphericity Statistical Analysis

In addition to the surface state, we have calculated another parameter describing the shape. This is the sphericity of the quartz particles. This allowed us to interpret the evolution of the roundness of the particles through the analysis of SEM images. The basic results of statistical analysis of textures on 43 quartz grains are given in Table 1.3 and the distribution of roundness of the quartz particles is shown in Figure 1.8.

This parameter was evaluated on the basis of data from the analysis of digital photos of exoscopic observation of quartz grains taken from four samples of the Issen bed. The results obtained confirm and complete those of the morphoscopy. The results of analysis of variance, carried out to study the spatial variations of the roundness of the quartz particles, showed that at the significance level of 10%, and standard deviation (0.04 to 0.039), there are significant variations in parameters of particles among the different sites (Table 1.4). The lowest value of the mean was observed at sample E5 in the upstream, while highest value was observed at sample E24 in the downstream. The maximum value for sphericity was registered at sample E24, while the highest value of skewness was observed at sample E5. The minimum sphericity parameter was registered at sample E5. In fact, the low level of the parameters studied in this location were originating primarily from the detachment in the rock downstream (Figure 1.7).

The decrease of the roundness appears from sample E17 and continues downstream in E12 at more than 54 km. However, at E12 there is an increase in the roundness of the particles after the confluence with Ida-Ou-Mahmoud (E6). This tributary is a source of more

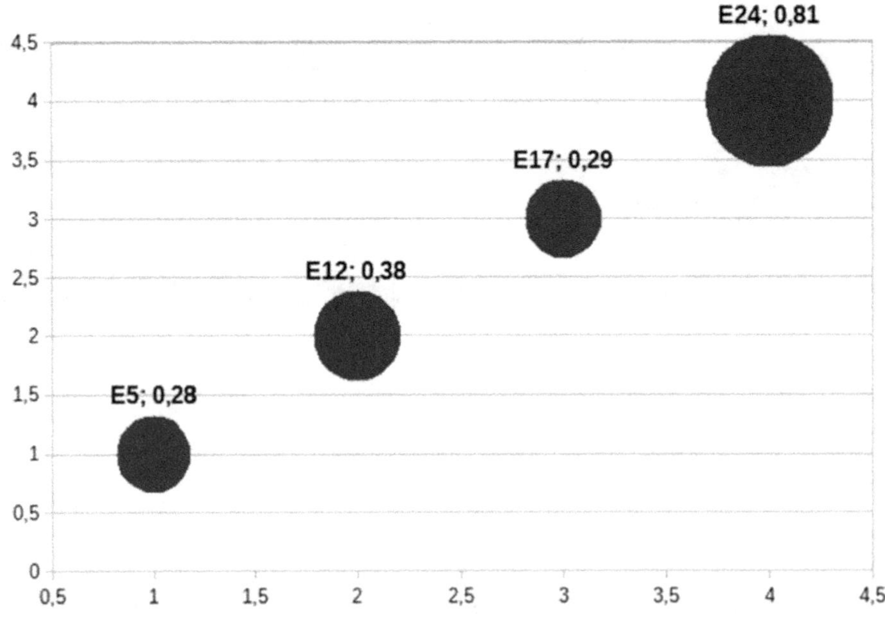

FIGURE 1.8 Evolution of the roundness of the quartz particles at different locations of the Issen Wadi.

TABLE 1.4 Statistics of Quartz Grain Sphericity Parameters of the Samples as Determined by SEM

Parameter's	E5	E12	E17	E24
Mean	0.28	0.38	0.29	0.81
Standard error	0,08	0,10	0,09	0,03
Mode	0,00	0,00	0,00	—
Median	0,19	0,44	0,18	0,81
First quartile	0,00	0,00	0,00	0,80
Third quartile	0,49	0,73	0,60	0,83
Variance	0,10	0,15	0,10	0,00
Standard deviation	0,31	0.39	0.32	0.04
Kurtosis	−1.70	−2.00	−2.08	—
Skewness	0.35	0.09	0.24	—
Range	0.78	0.92	0.70	0.06
Minimum	0.00	0.00	0.00	0.78
Maximum	0.78	0.92	0.70	0.84

or less angular particles that disrupt the longitudinal trend of the river towards particle size reduction and grain rounding.

1.5 CONCLUSION

The microtextures on the surface of quartz grains provides an insight into the sedimentary history of sediments and specify the nature of the transport agents causing their shaping. The study area – the Issen basin in the western High Atlas in Morocco – represents a particularly appropriate testing ground for research for the purpose of this work. This is due to the high efficiency of tectonics, lithology varied terrain relief, conditioned mainly by topographical, geomorphological and climatic processes. In this chapter, we have described the sedimentological parameters measured (calcemetery of the Oued Issen deposits and exoscopy of quartz grains). The surface of the quartz grains of the four samples analyzed is marked by numerous traces that have shapes and sizes characteristic of the factors that formed them. These factors can be of physical, chemical, mechanical, or biological origin. They showed significant spatial variations. The Issen sands essentially come from the erosion of the sandstone, areno-lutite and conglomeratic formations of the Triassic of the Argana corridor and secondarily from the schists and gewaeks of the primary massif. The present study has shown that the right bank tributaries draining the Jurassic plateaus are the most active in the distribution of limestone in the catchment. However, in general, the sediments are very poor in carbonates and fine sedimentation is developed due to significant water erosion.

REFERENCES

Ait Haddou, M., Bouchriti, Y., Ikirri, M., Wanaim, A., Aydda, A., Amarir, S., … & El Boudribili, Y.: Delineation of Groundwater Potential Zones in a Semi-arid Region Using Remote Sensing and GIS: A Case Study of Argana Corridor (Morocco). In *Advanced Technology for Smart Environment and Energy*, Cham: Springer International Publishing. pp. 257–268 (2023).

Ait Haddou, M., Kabbachi, B., Aydda, A., Bouchriti, Y., Gougueni, H., En-Naciry, M., Aichi, A.: Traditional practices: A window for water erosion management in the Argana basin (Western High Atlas Morocco). *E3S Web of Conferences*, 337, 02002 (2022a).

Ait Haddou, M., Kabbachi, B., Aydda, A., Gougni, H., Bouchriti, Y.: Spatial and temporal rainfall variability and erosivity: Case of the Issen watershed, SW-Morocco. *E3S Web of Conferences*, 183, 02003 (2020).

Ait Haddou, M. Wanaim, A. Ikirri, M. et al.: Digital elevation model-derived morphometric indices for physical characterization of the Issen basin (Western high Atlas of Morocco). *Ecological Engineering & Environmental Technology*, 23(5), 285–298 (2022b).

Blott, S. J., & Pye, K.: GRADISTAT: A grain size distribution and statistics package for the analysis of unconsolidated sediments. *Earth Surface Processes and Landforms*, 26(11), 1237–1248 (2001).

Bouchriti, Y., Kabbachi, B., Sine, H. et al.: COVID-19 prevention and control interventions: What can we learn from the pandemic management experience in Morocco? *The International Journal of Health Planning and Management*, 37(3), 1827–1831 (2022).

Ed-Dakiri, S. et al.: Water Erosion Risks Mapping Using RUSLE Model in the Mohamed Ben Abdelkrim El Khattabi Dam Watershed (Central Coastal Rif, Morocco). In: Dubey, S. K., Jha, P. K., Gupta, P. K., Nanda, A., Gupta, V. (eds) *Soil-Water, Agriculture, and Cli-mate Change*. Water Science and Technology Library, 113 (2022).

Elmouden A., Bouchaou L., Snoussi M., Wildi W.: Comportement des métaux et fonctionnement d'un estuaire en zone sub aride: cas de l'estuaire du Souss (côte atlantique marocaine) *Estudios Geológicos*, 61: 25–31 (2005).

Flemming, B. W.: Depositional Processes in Saldanha Bay and Langebaan Lagoon (Western Cape, South Africa). (Doctoral dissertation, University of Cape Town). (2015).

Fournier, J., Bonnot-Courtois, C., Paris, R., Voldoire, O., & Le Vot, M.: *Analyses granulométriques, principes et méthodes*. Dinard: CNRS, 100 p. (2012).

Lee, C. H., Li, L., Gong, J.: Hybrid higher-order skin-topological modes in nonreciprocal systems. *Physical Review Letters*, 123(1), 016805 (2019).

Mahaney, W. C.: *Atlas of sand grain surface textures and applications*. Oxford University Press, USA (2002).

Mondal, A., Khare, D., Kundu, S., Meena, P. K., Mishra, P. K., & Shukla, R.: Impact of climate change on future soil erosion in different slope, land use, and soil-type conditions in a part of the Narmada River Basin, India. *Journal of Hydrologic Engineering*, 20(6), C5014003 (2015).

Omdi, F. E., Daoudi, L., & Fagel, N.: Origin and distribution of clay minerals of soils in semi-arid zones: Example of Ksob watershed (Western High Atlas, Morocco). *Applied Clay Science*, 163, 81–91 (2018).

Rollet, A.J.: Etude et gestion de la dynamique sédimentaire d'un tronçon fluvial à l'aval d'un barrage: le cas de la basse vallée de l'Ain. *Géomorphologie. Université Jean Moulin - Lyon* III (2007).

Shepard, F. P.: Nomenclature based on sand–silt–clay ratios. *Journal of Sedimentary Petrology*, 24, 151 (1954).

Soyer, J.: Aspects de surface de sables quartzeux au microscope électronique à balayage. Annales de la Société géologique de Belgique [En ligne], Volume 92, *Fascicule*, 2, 215–227 (1969).

Tixeront, M.: Carte géologique et minéralisations du Couloir d'Argana. *Notes Mém. Serv. Géol. Maroc*, 1974; 205 (1974).

Vörösmarty, C. J., Meybeck, M., Fekete, B., Sharma, K., Green, P., & Syvitski, J. P.: Anthropogenic sediment retention: major global impact from registered river impoundments. *Global and Planetary Change*, 39(1–2), 169–190 (2003).

Vos, K., Vandenberghe, N., & Elsen, J.: Surface textural analysis of quartz grains by scanning electron microscopy (SEM): From sample preparation to environmental interpretation. *Earth-Science Reviews*, 128, 93–104 (2014).

Zhiying, Li, & Fang, H.: Impacts of climate change on water erosion. A review. *Earth-Science Reviews*, 163, 94–117 (2016).

CHAPTER 2

Impact of Agricultural Soil Degradation on Water and Food Security

Ahmed Elshaikh

University of Khartoum, Khartoum, Sudan

Jamal Mabrouki

University Mohammed V in Rabat, Morocco

2.1 INTRODUCTION

Soils in good conditions are necessary to support agricultural, biological, and environmental systems, and threats to soil resources are considered threats to human security itself (Amundson et al. 2015). The rapid population growth rates and the high consumption rates of individuals have led to an increase in demand for agricultural products, which places heavy pressure on natural resources, contributing to the disruption of ecosystems and their degradation in various forms (FAO 2017). The most important aspect of the future population dynamics is the fact that the majority of the population increase will be in developing countries (in South Asia and sub-Saharan Africa). In these regions, healthy soils are limited, vulnerable to natural and anthropogenic variables, and degraded through the increase in demographic pressure and the hazards of climate change. Thus, any future increase in food production will have to occur through vertical increase through the resources already committed to agriculture production per unit area in terms of nutrients, water, energy, etc. (Lal 2008). From many factors threatening soils, unsustainable agricultural production is arguably one of the largest parameters (Kopittke et al. 2019). Efforts conducted to improve the sustainability of agricultural soil must not only incorporate biological and agronomic knowledge, but also consider the human dimension in the agricultural production process (Lal 2008). Recently, there has been a renewed effort to analyze the economic dimensions of soil health in the agricultural context and to better integrate insights from the social sciences into ongoing policy discussions about soil management

DOI: 10.1201/9781003436218-2

(Kik et al. 2021; Rejesus et al. 2021). These include the exploitation of fragile and unstable environments in which the dynamic balance between their components is irreversible, or in a way that does not allow for great flexibility in dealing and responding to reasonable investment methods (FAO 2017).

2.2 SOIL DEGRADATION

Soil degradation could be defined in broad terms as the functional, physical, chemical, and biological decline in soil quality. This may appear in terms of loss of soil biodiversity, a decline in soil fertility (organic matter and structural condition), erosion, adverse changes in salinity, acidity or alkalinity, and the effects of toxic chemicals, pollutants, or excessive flooding (Nunes et al. 2020).

Moreover, soil degradation causes include unsustainable agricultural practices; overgrazing; agricultural, industrial, and natural pollution; and long-term impacts of climatic change (Maximillian et al. 2019).

Examples of soil degradation on the ground are shown in the follow Figures 2.1–2.6:

- **Water erosion**
- **Wind erosion**
- **Structure decline**
- **Soil mass movement**
- **Waterlogging**
- **Salinity**

FIGURE 2.1 High erosion rates after a thunderstorm on two-year-old apricot orchards (Keesstra et al. 2016).

FIGURE 2.2 Primary processes of wind erosion with photo examples of each (Tatarko et al. 2019).

FIGURE 2.3 Soil erosion in the Central African Republic (https://desertification.wordpress.com/2015/02/24/soil-fertility/).

FIGURE 2.4 Substrate mass movement processes (Stokes et al. 2013).

FIGURE 2.5 Sub-surface water table (previously drained), Ireland (Photo Dr. David Wilson).

FIGURE 2.6 Area with soil sealing and salinization. Municipality of Bom Jesus da Lapa, Bahia, Brazil (Nunes et al. 2020).

FIGURE 2.7 Relationship between soil degradation and water; agriculture; and human activity and the ecosystem.

Declining soil quality has significant implications for the development of future agricultural and water resources. Furthermore, it has great impacts on human health and ecosystem services. Figure 2.7 illustrates the relationship between soil degradation and water; agriculture; and human activity and the ecosystem.

2.3 HISTORICAL EVIDENCE

In Mesopotamia during the Bronze Age (4000–1100 BC) urban centers were focused in the cities of Sumer and Acad (Figure 2.8). The population of these cities was between 50,000 and 80,000, and about 90% of the population lived in such areas. At the beginning of the development of these kingdoms, almost half of the population worked in the large irrigated

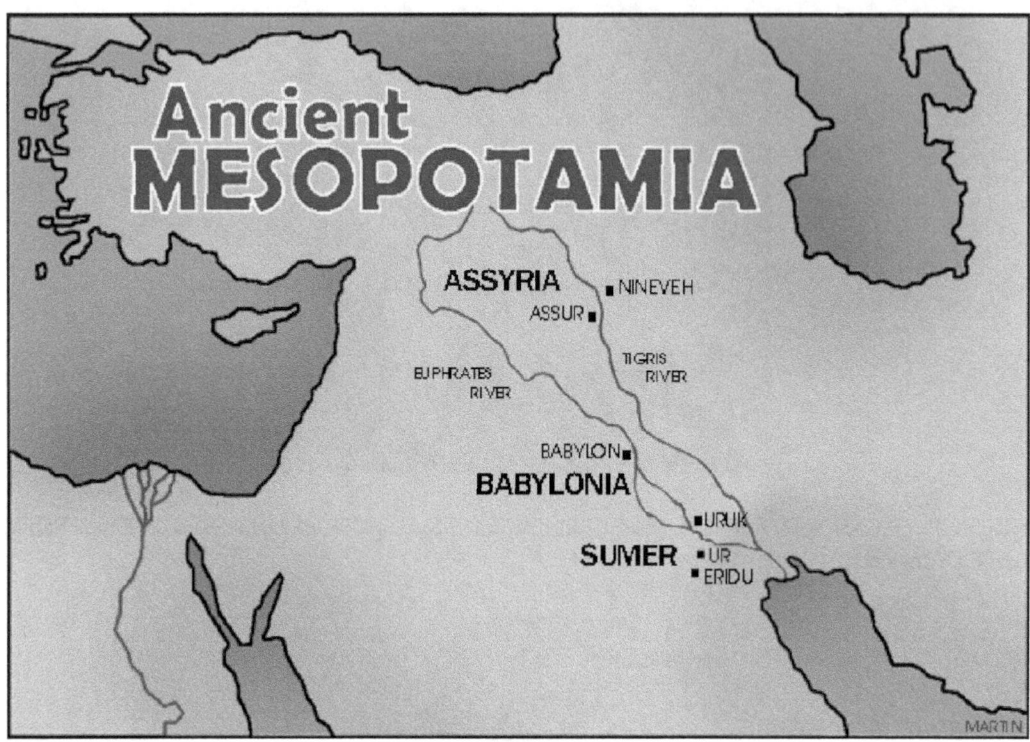

FIGURE 2.8 Location of Ancient Kingdoms in Mesopotamia (Donn 2019).

farms surrounding each city. As modern methods and tools developed during the time of the Sumerians, Mesopotamia witnessed major development in irrigated agriculture, with just 1% of the community able to feed the rest of the population. Encouraging farmers to focus on effective water management, in one panel, it was stated that the yield could be increased by 10% with optimal irrigation management (Mays 2010). Ancient farmers also used various means to irrigate their land, including Shadouf, waterwheel (Noira), irrigation canals, and dams.

One of the main causes for the deterioration of Sumerian civilization was the decline in agricultural production. Some pieces of evidence were found in cuneiform records indicating the extent of the salinization of agricultural soil. It was reported, for example, that the wine production of cereals in 2400 BC was 2347 liters per hectare, but the level of production had dropped to 897 liters per hectare in 1700 BC. Another indicator of soil salinization is the high proportion of barley areas, to more than 98% of the total grain planted in 2050 BC. Barley is known to be a salt-tolerant crop and is often the first crop used to restore saline desert soils for agriculture. The problem of silt deposition in many channels was widespread and influential, as it was easier to abandon some of these channels and construct new ones (Mays 2010).

2.4 SOIL DEGRADATION AND WATER SECURITY

Of all the major human activities, agriculture remains the largest user of water at the global level, accounting for about 70% of total freshwater withdrawal (Hillel et al. 2008).

Moreover, the agricultural sector is the largest source of freshwater withdrawal and water is a major constraint to food production in most irrigated projects all over the world.

Whereas agriculture is the largest consumer of water, the pressure on water resources will be exacerbated by soil degradation and severe drought stress resulting from the projected climate change. Water scarcity is aggravated by soil challenges and non-point and point source pollutions (Lal 2008). Water and soil relation depends upon different parameters, such as the rate of infiltration into the soil and on soil hydraulic conductivity that redistributes water through the soil. Similarly, water content and transmission times are also important to the function of soil because the contact with soil surfaces and residence time in soil are considered to be contamination controls (Montanarella et al. 2015). Figure 2.9 describes the different roles of soil and their impacts on water availability.

In dry regions, inadequate soil moisture can be mitigated through supplementary irrigation, and where excessive precipitation causes problems, waterlogging can be relieved by land drainage. Irrigation that enables a shift to intensive land use can increase the contaminant load of runoff and drainage water. However, irrigation and drainage can have consequences for water regulation services (McDowell et al. 2015).

Furthermore, an appropriate drainage system of soil will reduce water and contaminant storage capacity in the landscape; however, it will increase the potential floods downstream. Irrigation of agricultural lands accounts for about 70% of ground and surface water withdrawals, and in some regions competition for water resources is forcing irrigators to tap unsustainable sources (Montanarella et al. 2015). The abstraction of surface and groundwater for irrigation introduces some changes to the natural water cycle and may stress downstream ecosystems and communities. This will affect water security in terms of water availability and a great amount of water consumption to deal with soil degradation challenges (Figure 2.10).

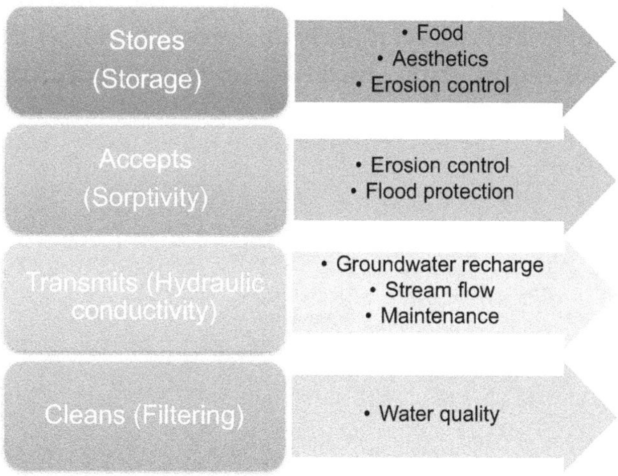

FIGURE 2.9 Soil and the water cycle (Montanarella et al. 2015).

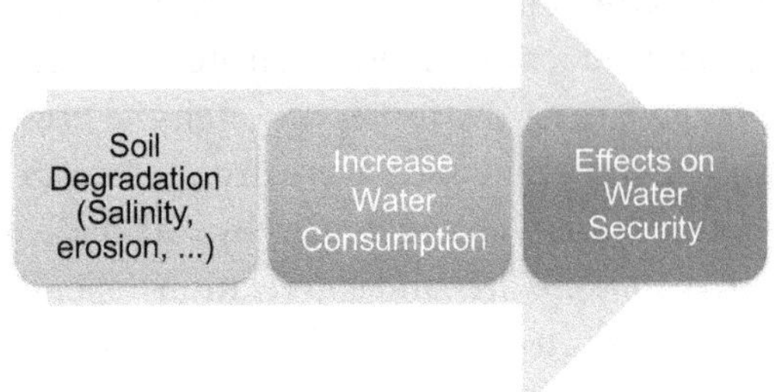

FIGURE 2.10 Soil degradation and water security.

2.5 SOIL DEGRADATION AND FOOD SECURITY

The Green Revolution is needed to feed the growing global population, and this must be based on sustainable management of soil and water resources (Lal 2009). Despite substantial investment in irrigation projects in the past decades, global irrigated lands are facing difficult situations. Irrigation plays a major role in providing food for the growing population, contributing over 40% to total agricultural production. Average cereal yields under irrigation are typically twice those obtained under rain-fed conditions (Hillel et al. 2008).

Soil degradation is one of the main factors affecting the sustainability of irrigated agriculture. Soil salinity results in the tendency to apply water in excess of irrigation requirements. This may also lead to waterlogging problems, which lead, in turn, to reduced aeration, nutrient uptake, and crop yields. Salinity hazards will increases if saline water is used for irrigation and when poor fertilizer and poor irrigation management are combined. All these forms of degradation directly affect the agricultural productivity and food production.

Furthermore, continuous cultivation of land results in low and declining crop yields and an inability to attend to other farm production constraints, and eventually to food deficits, low incomes, and food security challenges (Figure 2.11).

Because of the complex causes of low crop yields and their far-reaching effects, to ensure food security, no simple intervention is likely to overcome yield limitations, uplift households, and restore soil fertility. Rather than that, an integrated approach involving access

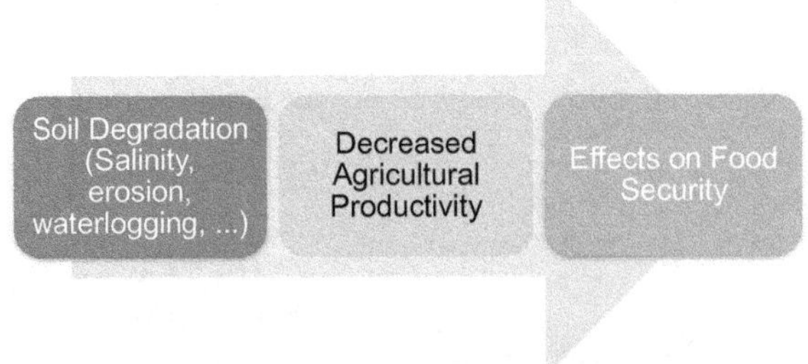

FIGURE 2.11 Soil degradation and food security.

to farm inputs, technologies to ensure their efficient use, land conservation measures, and improved social awareness (Bekunda et al. 2010).

2.6 SOIL DEGRADATION AND ENVIRONMENTAL IMPACTS

Soil degradation has important impacts on ecosystem services, and poor soil quality may diminish the ability of ecosystems to support the functions or services required to ensure their sustainability. Growing awareness about the interdependence of parts of ecosystems will put more pressure on agriculture to reduce negative impacts on ecosystems, for example, by reducing soil degradation and managing carbon storage. In addition, soil degradation represents a serious threat to environmental issues that are concerned with land stability and cover (Gobinath et al. 2022).

Land-use change can significantly reduce soil organic carbon (SOC) and increase carbon dioxide (CO_2), nitrous dioxide (NO_2), and methane (CH_4) emissions. Changing the use of land from pasture to cultivation area results in the greatest loss of soil organic carbon. Soil organic matter plays an important role in creating soil organic carbon. In drought regions and during fire events, soil carbon is returned to the atmosphere as carbon dioxide, which affects directly the global warming and climate change process (Lehmann et al., 2015). Due to limited land and water resources, environmental planning will increasingly constrain the release of resources for cultivation purposes (Fischer et al. 2010; Molden et al. 2007).

2.7 PREVENTING AND RESTORING DEGRADED SOILS

Protecting and restoring the quality of degraded soils is a challenging task; the restoration of soil quality is a complicated task that necessitates a coordinated approach. There are many basic procedures to prevent land degradation that are described as follows (Dragović and Vulević 2020):

- Human-induced degradation develops more rapidly than natural degradation and can be reduced or avoided by regulating human interventions such as deforestation, overgrazing, and the mismanagement of agricultural land.
- In order to prevent soil degradation, it is necessary to monitor and assess soil degradation processes using appropriate methods (for example, expert-based, remote sensing, modeling...).
- It is necessary to close data gaps, enable access to data and data comparability, gather more on-the-ground information, and take into account the uncertainty of the future.
- Soil erosion is the most widespread form of soil degradation. Soil erosion control measures include the selection of optimal land use, the maintenance of a protective cover (trees, mulches, and crops), and the construction of technical works (e.g., dams).
- Soil contamination is the chemical degradation of soil that involves the presence of harmful substances in the soil. Remediation choices include a variety of chemical and physical treatments of soil, based on the degradation conditions.

- Salinization is a widely present type of chemical soil degradation that occurs in both arid and semi-arid areas. The expansion of salt-affected soil could be reduced by applying appropriate irrigation practices, the better use of fertilizers, or changes in land use.
- Degradation-type identification is important to define all consequences and, thus, the expected required actions and cost of restoration.
- It is essential to be aware that several degradation processes may occur simultaneously in the same area.
- The spatial extent, degree, and rate of degradation type should be a base for the decision-making process regarding the best sustainable land management practice.

Considering the technological advances that have been made in recent years, most types of soil degradation are potentially reversible, as long as there is sufficient public support, understanding, and political will (Lal 2015).

To achieve the target goal, some strategies were developed for soil quality restoration, which includes: (i) creating a positive soil/ecosystem; (ii) reducing soil erosion; (iii) improving the availability of fertilizers; and (iv) increasing soil biodiversity especially the microbial processes (Lal 2015). The ultimate goal should be to adopt a holistic and integrated approach to soil resource management (Figure 2.12).

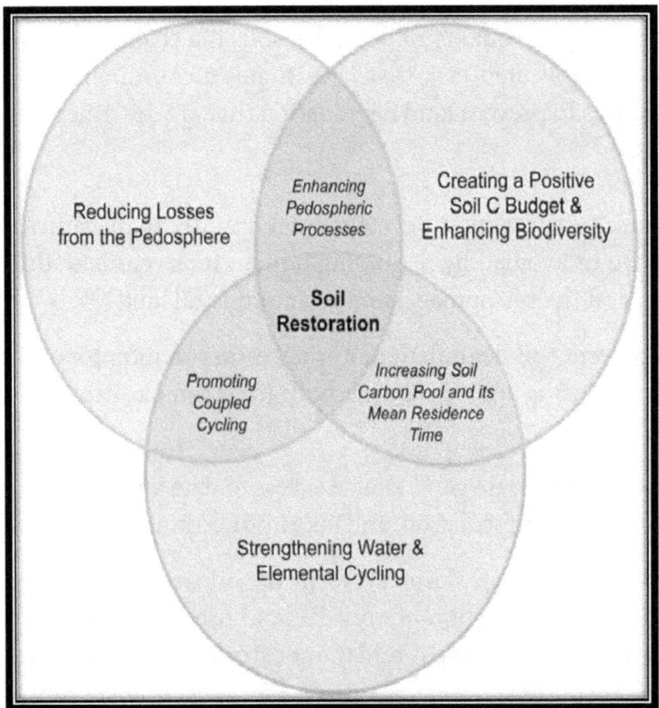

FIGURE 2.12 Three strategies of restoring and managing soil quality for mitigating risks of soil degradation (Lal 2015).

2.8 CONCLUSION

Soil degradation is a serious threat to water resources and sustainable agriculture. It could occur as a natural process or cause by human intervention. The mismanagement practice in agriculture led to soil degradation in terms of salinity and nutrient transportation, which in turn caused an increase in water demand and a decrease in land productivity. The excess irrigation water puts more pressure on the available water resources, especially in water scarcity regions. To emerge from this vicious cycle, it is extremely important to adopt strategies and techniques that aim to prevent and restore degraded soil to ensure soil sustainability. Moreover, under climate change scenarios of increased climatic variability with more extremes of precipitation, soil functions will be stressed and better soil management will be required to achieve water and food security.

Recommendations

Sustainable soil management is a key solution for soil degradation. In this regard, an improved understanding of soils and agronomic processes which enhance water use efficiencies is extremely critical.

Long-term actions include improving agriculture and water management practices, such as adopting proper crop rotation, enhancing soil drainage, improving irrigation efficiencies, using crops which are suitable for the climate, and capacity building for farmers. Farming practices can be employed to improve soil quality and increase soil carbon, including optimal fertilization and animal manure application. Efficient water management, along with better fertilizer use and improved crop varieties, could significantly reduce the negative effects of agricultural soil degradation. Finally, soil health must be given a central position in decisions made to combat climate change.

REFERENCES

Amundson, R., Berhe, A. A., Hopmans, J. W., Olson, C., Sztein, A. E., and Sparks, D. L. (2015). "Soil and human security in the 21st century." *Science*, 348(6235), 1261071.

Bekunda, M., Sanginga, N., and Woomer, P. L. (2010). "Restoring soil fertility in sub-Sahara Africa." *Advances in Agronomy*, 108, 183–236.

Donn. (2019). "Ancient Mesopotamia." https://mesopotamia.mrdonn.org/geography.html (Nov. 3, 2019).

Dragović, N., and Vulević, T. (2020). "Soil Degradation Processes, Causes, and Assessment Approaches." In *Life on Land*, Springer, 928–939.

FAO. (2017). *The Future of Food and Agriculture – Trends and Challenges*. FAO, Rome.

Fischer, M., Bossdorf, O., Gockel, S., Hänsel, F., Hemp, A., Hessenmöller, D., Korte, G., Nieschulze, J., Pfeiffer, S., and Prati, D. (2010). "Implementing large-scale and long-term functional biodiversity research: The biodiversity exploratories." *Basic and Applied Ecology*, 11(6), 473–485.

Gobinath, R., Ganapathy, G. P., Gayathiri, E., Salunkhe, A. A., and Pourghasemi, H. R. (2022). "Ecoengineering practices for soil degradation protection of vulnerable hill slopes." *Computers in Earth and Environmental Sciences*, 255–270.

Hillel, D., Braimoh, A. K., and Vlek, P. L. G. (2008). "Soil Degradation Under Irrigation." *Land Use and Soil Resources*, edited by Hamid Reza Pourghasemi, Springer, 101–119. Amsterdam: Elsevier.

Keesstra, S., Pereira, P., Novara, A., Brevik, E. C., Azorin-Molina, C., Parras-Alcántara, L., Jordán, A., and Cerdà, A. (2016). "Effects of soil management techniques on soil water erosion in apricot orchards." *Science of the Total Environment*, 551, 357–366.

Kik, M. C., Claassen, G. D. H., Meuwissen, M. P. M., Smit, A. B., and Saatkamp, H. W. (2021). "The economic value of sustainable soil management in arable farming systems–A conceptual framework." *European Journal of Agronomy*, 129, 126334.

Kopittke, P. M., Menzies, N. W., Wang, P., McKenna, B. A., and Lombi, E. (2019). "Soil and the intensification of agriculture for global food security." *Environment International*, 132, 105078.

Lal, R. (2008). "Soils and sustainable agriculture. A review." *Agronomy for Sustainable Development*, 28(1), 57–64.

Lal, R. (2009). "Soils and world food security." *Tillage Research*, 102(1), 1–4.

Lal, R. (2015). "Restoring soil quality to mitigate soil degradation." *Sustainability*, 7(5), 5875–5895.

Lehmann, J. R. K., Nieberding, F., Prinz, T., et al. (2015). "Analysis of unmanned aerial system-based CIR images in forestry—A new perspective to monitor pest infestation levels." *Forests*, 6(3), 594–612.

Maximillian, J., Brusseau, M. L., Glenn, E. P., and Matthias, A. D. (2019). "Pollution and Environmental Perturbations in the Global System." *Environmental and Pollution Science*, Elsevier, 457–476.

Mays, L. (2010). *Ancient Water Technologies*. Springer Science & Business Media.

McDowell, R. W., Cox, N., Daughney, C. J., Wheeler, D., and Moreau, M. (2015). "A national assessment of the potential linkage between soil, and surface and groundwater concentrations of phosphorus." *JAWRA Journal of the American Water Resources Association*, 51(4), 992–1002.

Molden, D., Oweis, T. Y., Pasquale, S., Kijne, J. W., Hanjra, M. A., Bindraban, P. S., Bouman, B. A. M., Cook, S., Erenstein, O., and Farahani, H. (2007). Pathways for increasing agricultural water productivity. *Water for Food Water for Life: A Comprehensive Assessment of Water Management in Agriculture*, pp. 279–314, Taylor & Francis.

Montanarella, L., Badraoui, M., Chude, V., Costa, I., Mamo, T., Yemefack, M., Aulang, M. S., Yagi, K., Hong, S. Y., and Vijarnsorn, P. (2015). "Status of the world's soil resources: main report." *Embrapa Solos-Livro científico (ALICE)*, FAO, 2015.

Nunes, F. C., de Jesus Alves, L., de Carvalho, C. C. N., Gross, E., de Marchi Soares, T., and Prasad, M. N. V. (2020). "Soil as a Complex Ecological System for Meeting Food and Nutritional Security." *Climate Change and Soil Interactions*, Elsevier, 229–269.

Rejesus, R. M., Aglasan, S., Knight, L. G., Cavigelli, M. A., Dell, C. J., Lane, E. D., and Hollinger, D. Y. (2021). "Economic dimensions of soil health practices that sequester carbon: Promising research directions." *Journal of Soil and Water Conservation*, 76(3), 55A–60A.

Stokes, A., Raymond, P., Polster, D., and Mitchell, S. J. (2013). "Engineering the ecological mitigation of hillslope stability research into the scientific literature." *Ecological Engineering*, 61, 615–620.

Tatarko, J., Trujillo, W., and Schipanski, M. (2019). *Wind Erosion Processes and Control*. Fort Collins, CO: Colorado State University Extencion.

CHAPTER 3

Effect of Olive Pomace Compost Application on the Growth and Yield of Fava Bean Plant

Halima Ameziane

Ibn Tofail University, Kenitra, Morocco

Mohammed V University, Sale, Morocco

Driss Hmouni

Ibn Tofail University, Kenitra, Morocco

Fatima Benradi and Abderrahman Nounah

Mohammed V University, Sale, Morocco

3.1 INTRODUCTION

The use of chemical fertilizers for soil amendment and tillage is a worldwide practice aimed at improving or protecting agricultural productivity. However, excessive or inappropriate use of these chemical products can have harmful effects on human health, wildlife, and the environment (Comoretto et al., 2007; Lanfer-Marquez et al., 2005). Organic amendments are encouraged to enhance the sustainability of agricultural systems, especially in arid and semi-arid areas where soils are often very poor in organic matter. The addition of organic amendments results in significant changes in the biological, chemical, and physical properties of soil, and these changes can influence the mobility and persistence of herbicides and thus modify their fate in the environment.

Indeed, the agri-food industry produces a wide variety of organic waste that can potentially be used as fertilizers and amendments for soils due to their high content of organic matter (OM) and plant nutrients (Martínez-Blanco et al., 2011). A clear example of this

DOI: 10.1201/9781003436218-3

issue is the olive oil industry, which has significant economic and social importance in Mediterranean countries and generates a huge amount of waste, depending on the extraction system used.

In Morocco, the world's eighth-largest producer, the agro-industry sector generates large amounts of waste and organic by-products during the short olive harvest season, whose progressive accumulation or incorrect disposal can have adverse effects on the environment. Among the by-products of the olive oil extraction process are olive pomace, a wet solid material obtained by continuous two-phase or three-phase centrifugation, or by traditional extraction (Ameziane et al., 2018). For its proper composting, the addition of several bulking agents, such as cattle manure (Ameziane et al., 2020b), shredded olive prunings (Regni et al., 2017; Proietti et al., 2015; Del Buono et al., 2011), grape stems (Baeta-Hall et al., 2005), sawdust and shavings, and poplar bark (Filippi et al., 2002), has been carried out to improve its low porosity; in all these cases, final products with high content of organic matter and significant amounts of nutrients for plants have been obtained.

However, it is increasingly evident that the application of raw or composted olive pomace to soils can have beneficial effects on soil properties and crop productivity (López-Piñeiro et al., 2008; Nicholls and Altieri, 2014). Similarly, the use of olive pomace has led to an improvement in soil biological activity (Innangi et al., 2017). The amendment with olive pomace has moved beyond olive groves to be used in herbaceous crop cultivation (Del Buono et al., 2011; Alburquerque et al., 2007; Brunetti et al., 2005; Tajada and Gonzalez, 2004). However, the abusive use of raw pomace as an amendment has shown some problems related to the time required for germination and root development due to its high organic load, mineral salts, low pH, and the presence of phytotoxic compounds (Proietti et al., 2015; Gigliotti et al., 2012; Del Buono et al., 2011). Therefore, composting is necessary to stabilize the high organic load and benefit from a soil amendment rich in humic compounds, cation exchange capacity, and water retention capacity (Mekki et al., 2013).

The objective of this work is to study the effectiveness of a compost made from a mixture of olive pomace and cattle manure, for use as soil amendment during the germination of fava bean seeds. Different concentrations of compost (5%, 10%, 15%, 20%, and 25%) are tested during the germination trial to determine the optimal dose. During the growth period, several parameters are regularly measured, including stem height, number of leaves, leaf surface area, number and size of pods. After seed maturity, they are harvested and subjected to a qualitative analysis of their nutritional value, measuring their protein, lipid, total sugar, and fiber content.

3.2 MATERIALS AND METHODS

3.2.1 Experimental Design

The experiment aims to germinate fava bean seeds in moderate percentages of compost made from olive pomace (Table 3.1), in order to identify the fertilizing power of this compost and determine the optimal dose useful for the plants. To do so, fava bean seeds are disinfected with bleach water, thoroughly washed with water, and then rinsed with distilled water. They are then germinated on a soil surface of 225 m^2 (15*15).

The germination site is located at the Higher School of Technology in Salé (Salé, Morocco, latitude 34°03′11″ North, longitude 6°47′54″ West, altitude above sea level: 34 m). The germination area studied is divided into six plots. Five plots contain a well-determined amount of compost corresponding to a defined percentage, 5%, 10%, 15%, 20%, and 25% compost, and the sixth plot is reserved for the control which only contains soil, whose characteristics are presented in Table 3.1. In each plot, fava bean seeds are sown (at a rate of 10 seeds/1.5 m^2) to a depth of 5 cm with a spacing of 10 cm between seeds.

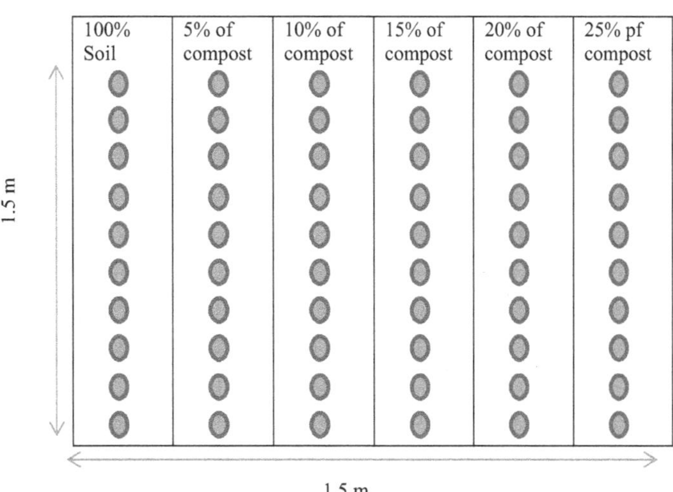

The physicochemical characteristics of the soil are presented in Table 3.1.

TABLE 3.1 Soil Physicochemical Parameters (Ameziane et al., 2020a)

Physicochemical Parameters	**Soil Characteristics**
Clay in %	9 ± 0.044
Sand in %	68 ± 0.088
Silt in %	23 ± 0.044
pH	7.61 ± 0.044
Electrical conductivity in μS/cm	85.5 ± 0.149
Moisture in %	4.51 ± 0.044
Organic matter in %	6.88 ± 0.125
Total organic carbon %	4.88 ± 0.333
Total nitrogen in %	0.134 ± 0.02
Total Phosphorus %	0.0103 ± 0.003
Calcium %	0.48
Potassium %	0.0445
Magnesium %	0.0243
Sodium %	0.0092

The values obtained represent the average of three repetitions.

TABLE 3.2 Physicochemical Properties and Germination Index of Mature Compost (Ameziane et al., 2020b)

Measured Parameters	Mature Compost
pH	8.61 ± 0.03
Humidity in %	30.4 ± 0.14
EC in mS cm^{-1}	2.06 ± 0.22
OM %	38.4 ± 0.76
K_2O %	2.8 ± 0.36
NTK %	1.3 ± 1.89
P_2O_5 %	0.42 ± 0.86
COT %	22.32 ± 0.89
C/N %	17.16
The germination index in %	73 ± 0.65

Note: The values obtained represent the average of three repetitions

TABLE 3.3 Physicochemical Parameters of Irrigation Water (Ameziane et al., 2020a)

Physicochemical Parameters	Characteristics of Irrigation Water
pH	7.49 ± 0.047
Electrical conductivity in mS/cm	1.024 ± 0.016
Temperature in °C	18.13 ± 0.058
Suspended matter in mg/l	0.232 ± 0.033
Salinity in ppb	0.3 ± 0.1
Nitrates in mg/l	7.04 ± 0.007
Chlorides in mg/l	192.5 ± 0.044
Boron in mg/l	0
Sulphates in mg/l	21.33 ± 0.005
Orthophosphates in mg/l	0.0134 ± 0.0001
The SAR	0.53 ± 0.5

The compost used in this study is the result of recent work (Ameziane et al., 2020b). It consists of 43% olive pomace and 57% cattle manure, which are mixed in large 30-liter barrels. The barrels are perforated to ensure an aerated environment and placed in a sunny location. Composting lasted 130 days to obtain mature compost, whose physicochemical characteristics are presented in Table 3.2.

Each plot was irrigated with 2 liters of water on the day of sowing, using water from the well of the Higher School of Technology in Salé, whose physicochemical characteristics are in compliance with the Moroccan standard for irrigation water (Secretariat of State at the Ministry of Energy, Mines, Water and Environment, in charge of Water and Environment, 2007) (Table 3.3). The plants were then left to the rainfall, with an average density of 550 mm/year.

a. Plant material: The broad bean

   The species Vicia faba is found in all regions of Morocco, and it includes several varieties, including Aguadulce, which was used in the present study. Aguadulce is a local variety that is sensitive to water scarcity and prefers fresh, deep, and slightly acidic soils (Sibennasseur Alaoui, 2005). It is sensitive to soil compaction and excess water, but it is resistant to cold temperatures (Sibennasseur Alaoui, 2005).

b. Measured parameters
Regular monitoring of the broad bean plant was carried out. Five parameters were determined:

- Height of aboveground parts (stems): The height of the stem was measured using a graduated ruler, from the soil surface to the tip of the leaves. Several measurements were taken throughout the experiment.
- Number of leaves: The number of leaves was counted for each plant.
- Leaf surface area: The leaf surface area was calculated by multiplying the length (*L*) by the width (*l*) of each leaf (Parcevaux and Catsky, 1970)

$$S = L \times l \tag{3.1}$$

- Number of pods: After harvest, the number of pods contained in each plant was counted.
- Pod size: The size of the pods was measured using a graduated tape measure.

c. Nutritional quality
At the end of the germination test, and for each percentage, the broad bean seeds contained in the pods were collected, dried at 60°C until reaching a constant weight, then ground and sieved to 0.2 mm to determine their protein, lipid, total sugar, and dietary fiber content. The assessment of the nutritional quality was carried out in the laboratory of the Regional Center for Agricultural Research in Rabat, Food Technology Research Unit.

i. Protein determination: Kjeldahl method (Bremner and Mulvaney, 1982)
Proteins are determined by the Kjeldahl method, which involves the hot decomposition, with sulfuric acid and a catalyst, of organic compounds containing nitrogen (proteins and nucleic acids). This catalyst contains potassium sulfate (K_2SO_4), which increases the boiling point of sulfuric acid, and copper sulfate ($CuSO_4$), which acts as a catalyst for the reaction. Nitrogen is quantitatively converted into ammonium sulfate during the mineralization step.

Ammonia is then displaced from its salt by sodium hydroxide, distilled with steam, and collected in a known amount of excess hydrochloric acid. This is the distillation step. The amount of hydrochloric acid that did not react is then titrated with sodium hydroxide. This is the titration step. The nitrogen content is expressed by the formula: (3.2)

$$\%MAT = \frac{\left(\left(V_e - V_b\right)N_{HCl} \times 14 \times 6{,}25\right)}{P_e \times 100} \tag{3.2}$$

With:

V_e: the volume of HCl used for titrating the sample in ml

V_b: the volume of HCl used for titrating the blank in ml

N_{HCl}: the normality of HCl in N

P_e: the weight of the sample in mg.

ii. Lipid determination: Soxhlet method
The sample is weighed and then placed in a cellulose cartridge. The cellulose cartridge is permeable to the solvent and the fat that is dissolved in it. The solvent in the heating flask rises to the extraction chamber for 5 to 10 minutes and then returns to the flask. The sample is extracted semi-continuously by boiling ethyl ether, which drips into the extraction chamber and gradually dissolves the fat. The ethyl ether containing the fat returns to the flask through successive spills caused by a siphon effect in the side elbow. As only the solvent can evaporate again, the fat accumulates in the flask until the extraction is complete. Once the extraction is complete, the solvent is evaporated in an oven at 100°C, and the lipid content is measured by the weight of the extracted lipids (AOAC, 1990).

iii. Total sugar determination: Dubois et al. (1956) phenol method
Total soluble sugars (sucrose, glucose, fructose, their methyl derivatives, and polysaccharides) are measured using the Dubois et al. (1956) phenol method, which involves concentrated sulfuric acid causing the release of several water molecules from the sugars when heated. This dehydration is accompanied by the formation of hydroxymethylfurfural in the case of hexose and furfural in the case of pentose. These compounds condense with phenol to form colored complexes (yellow-orange). The intensity of the color varies depending on the amount of sugars present in the reaction mixture. The absorbance is measured at a wavelength of 492 nm using a spectrophotometer.

iv. Determination of total fiber: NDF method
To determine the total fiber content, a neutral detergent solution (NDF) is used to dissolve proteins, lipids, sugars, easily digestible starches and pectins in food, leaving an insoluble fibrous residue that is mainly composed of cell wall components of plant materials (cellulose, hemicellulose, and lignin) and indigestible nitrogenous matter in the bean. The results are reported on a dry matter basis. (ISO 16472: 2006).

3.3 STATISTICAL ANALYSIS

The experimental data was subjected to a one-way analysis of variance (ANOVA) and mean separations were performed by the least significant difference (LSD) at a significance level of $P < 0.05$, using the Statgraphics Centurion XVI program for Windows.

3.4 RESULTS

3.4.1 Effect of Compost Addition on Stem Size

Compost addition to the fava bean plant improved stem size for the different percentages compared to the control (Figure 3.1). Indeed, the stem size of plants fertilized with 5%, 10%, 20%, and 25% is significantly greater than that of control plants. Stem growth is often linked to nitrogen abundance. Nitrogen is involved in several processes and structures in plants, such as the formation of amino acids, which form proteins inside the plant, necessary for its growth (Jake Munroe, 2018).

a. The effect of compost addition on the number of leaves
   The number of leaves of plants fertilized with different percentages of compost from olive pomace is significantly higher compared to the number of leaves of control plants (Figure 3.2). This high number of leaves can be explained by the richness of compost in nitrogen (1.3%) compared to the soil (0.134%) (Table 3.2). Indeed, nitrogen is the most important element in the formation and development of leaves. It plays an important role in the production of chlorophyll, responsible for converting sunlight into energy (through photosynthesis) that the plant needs (Mérigout, 2006). This results in green and robust leaves. A lack of nitrogen mainly manifests in leaves as yellowing (Bové, 1962).

b. Effect of compost addition on leaf surface area
   The addition of compost led to a significant improvement in the leaf area of the fava bean plants for all doses (5%, 10%, 15%, 20%, and 25%) compared to the leaf area of the control plants (Figure 3.3). As mentioned earlier, the development and growth of plant leaves are often linked to the abundance of nitrogen, the main constituent of chlorophyll and proteins (Mérigout, 2006). Therefore, the significant concentration of this element in the compost (1.3%) compared to the control soil (0.134%) (Table 3.2) can be a convincing explanation for the results obtained.

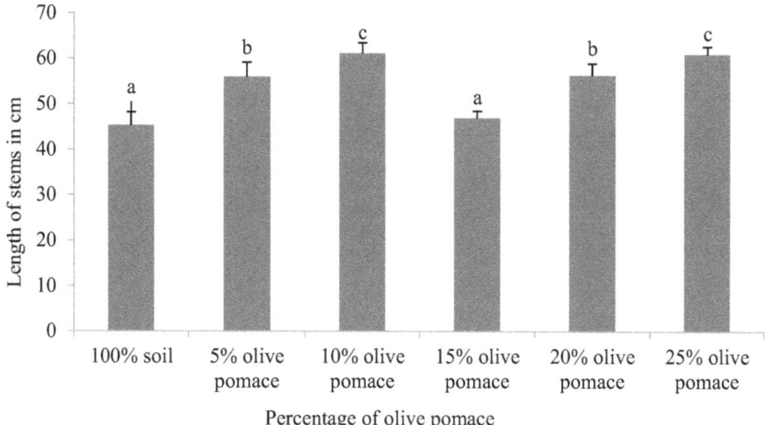

FIGURE 3.1 Effects of different concentrations of olive pomace on stem size of fava bean plant (values with different letters are significantly different: $p < 0.05$).

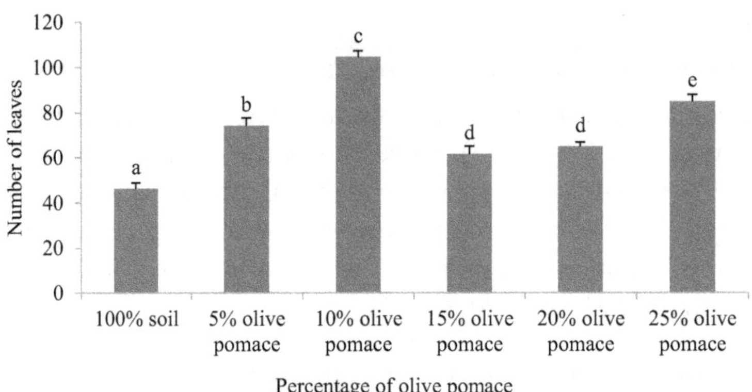

FIGURE 3.2 Effects of different concentrations of olive pomace on the number of leaves of fava bean plant (Values with different letters are significantly different: $p < 0.05$).

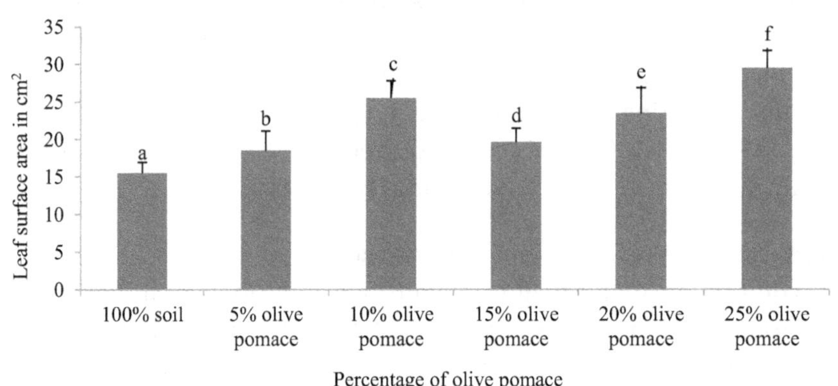

FIGURE 3.3 Effects of different concentrations of olive pomace on the leaf area of the fava bean plant (Values with different letters are significantly different: $p < 0.05$).

i. Effect of compost addition on the number and size of bean pods
The use of compost on the fava bean plant led to a significant increase in the production of fava bean fruits and a remarkable improvement in the size of the pods, which remained significantly different from the production and size of pods of control plants, for all doses used. The production and maturation of fruits are often linked to the richness of the soil in exchangeable potassium (Jake Munroe, 2018). Indeed, the studied compost presents appreciable levels of exchangeable potassium (2.8%) compared to the control soil (0.044%) (Table 3.2), which explains the high number and size of the obtained pods at fruiting. Similar results have been obtained by other authors (Clemente et al., 2012; Alburquerque et al., 2011; Walker and Bernal, 2008), who state that the high level of exchangeable potassium in olive pomace compost causes a significant increase in the availability of potassium in the soil. This increase is generally reflected by higher potassium concentrations in the leaves and fruits of plants grown in soils treated with olive pomace compost than in those receiving other amendments or no amendment (Martínez-Fernández and Walker, 2012).

TABLE 3.4 Nutritional Quality of Fava Bean Seeds

Soil Type/ Compost Dose	Proteins in g/100 g	Lipids in g/100 g	Total Sugars in g/100 g	Total Fibers in g/100 g
100% soil	$5.1^{a} \pm 0.02$	$0.8^{a} \pm 0.6$	$6.1^{a} \pm 0.65$	$3.1^{a} \pm 0.76$
5% compost	$7.5^{b} \pm 0.12$	$0.22^{b} \pm 0.98$	$7.95^{b} \pm 0.32$	$3.6^{a} \pm 0.45$
10% compost	$\mathbf{8.06^{b}} \pm 0.23$	$\mathbf{1.2^{a}} \pm 0.45$	$\mathbf{9.35^{c}} \pm 0.67$	$\mathbf{4.56^{b}} \pm 0.35$
15% compost	$5^{a} \pm 0.04$	$0.2^{b} \pm 0.92$	$4^{d} \pm 0.57$	$2.5^{a} \pm 0.67$
20% compost	$6.76^{c} \pm 0.32$	$0.67^{c} \pm 0.54$	$4.2^{d} \pm 0.45$	$5.85^{c} \pm 0.84$
25% compost	$6.04^{c} \pm 0.04$	$0.34^{b} \pm 0,84$	$4.7^{d} \pm 0.4$	$2.90^{a} \pm 0.73$

The values obtained represent the average of three repetitions; for each treatment, the averages in each column followed by the same letter are not significantly different at $P < 0.05$.
In each case, the figures in bold represent the compost dose at which the best result is achieved.

ii. The effect of adding compost on the nutritional quality of fava beans
Due to the high nutrient content of the compost, its addition has improved the nutritional quality of fava bean pods. A significant improvement was observed in terms of protein (8.06 g/100 g), lipid (1.2 g/100 g), total sugar (9.35 g/100 g), and fiber (4.56 g/100 g) concentrations for plants fertilized with 10% olive pomace compost compared to the concentrations of these elements in seeds germinated in the control (Table 3.4).

The improvement in the nutritional quality of fava (Table 3.4) beans is remarkable for the different percentages of compost, which can be explained by the richness of the compost in fertilizing elements compared to the soil, especially in major elements such as nitrogen, which plays an essential role in the formation of proteins and nucleic acids (Mérigout, 2006). Indeed, the protein content of the fruits is closely related to the abundance of nitrogen in the soil (Mérigout, 2006), as well as the richness of compost in phosphorus and potassium (Table 3.1 and 3.2), which stimulate root development, flowering, and fruiting by improving the synthesis of starch, sugars, and organic acids in the fruits, thereby improving their nutritional quality and flavor (Mérigout, 2006).

3.5 DISCUSSION

Composts made from olive pomace have been shown to be suitable organic fertilizers for horticultural crops (Alburquerque et al., 2006), olive trees (Cayuela et al., 2004), and also as a component of substrate or growing medium for ornamental plants (García-Gomez et al., 2002).

In terms of morphological results, it was observed that the addition of olive pomace compost increased the stem size, number of leaves, leaf surface area, as well as the number and size of pods compared to the control, for different compost doses (Figure 3.4). This morphological improvement of the fava bean can be attributed to the richness of olive pomace compost in fertilizing elements (NPK, P2O5, and K2O) necessary for the growth and development of most plants (Table 3.2) (Regni et al., 2016). Indeed, the addition of olive pomace compost to soils generally increases the total concentrations of hydrosoluble

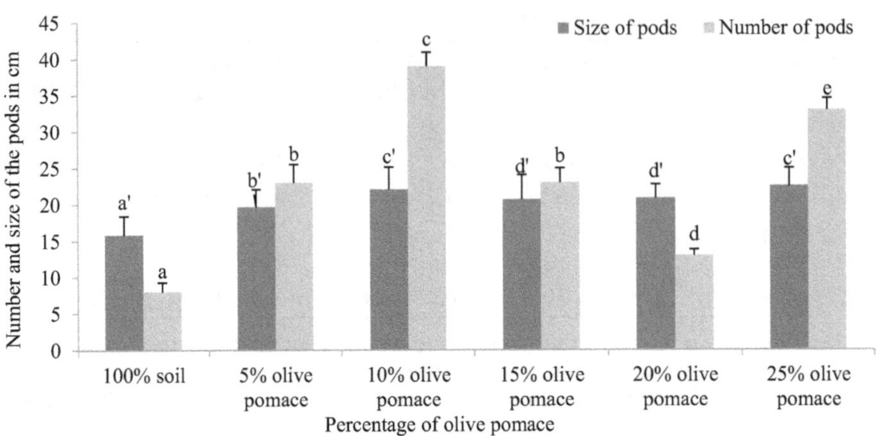

FIGURE 3.4 Effects of different concentrations of olive pomace on the size and number of bean pods (Values with different letters are significantly different: $p < 0.05$).

organic carbon as well as nitrogen, which mainly contributes to the development of plant foliage and aerial parts, potassium, which is useful for the circulation of sap and assimilation of nutrients by plants, and phosphorus, which stimulates root development, flowering, and fruiting (Cao et al., 2018; Curaqueo et al., 2014). Similarly, a study conducted by Clemente et al. (2012) and Pardo et al. (2014d) revealed a 10-fold increase in potassium and a 5-fold increase in phosphorus compared to the control in a soil that was poor in these two elements, after amending it for two years with olive pomace compost (Table 3.4). This improvement is of great importance for the stimulation of biogeochemical nutrient cycles in these typically poor soils and for their fertility (Pardo et al., 2014b).

In terms of nutritional quality, it was observed that the addition of compost at different doses significantly improved the nutritional quality of fava bean seeds compared to those of the control, especially for the 10% compost concentration. Similar results have been reported by other authors. For instance, during a two-year experiment, Tajada and Gonzalez (2004) observed an improvement in various characteristics of corn cultivation, such as crude protein content of the grain, soluble carbohydrate content in the grain, number of grains per ear, and yield, after applying olive pomace compost at doses of 10, 20, 30, and 40 t/ha. The research conducted by Alburquerque et al. (2007) also showed an improvement in plant growth through the application of compost made from olive pomace and cotton ginning wastes. Nasini et al. (2013) observed an improvement in fruit growth and yield per tree, without a reduction in oil content, following the application of olive pomace compost on olive trees. Similarly, over three consecutive years, Proietti et al. (2015) applied composted olive pomace to an olive grove, resulting in an increase in vegetative activity and productivity of the olive trees.

3.6 CONCLUSIONS

In conclusion, with regard to this experiment, the low fertility of the soil used and the significant content of major elements necessary for plant development (NTK, P2O5, K2O) present in the compost of olive pomace allowed for significant growth of the Fava bean plants. Adding different doses of compost (5%, 10%, 15%, 20%, and 25%) to the Fava bean plants improved their growth parameters. The number and leaf area of the plants were significantly higher for all doses used compared to the control group. In addition, the stem size of the plants was significantly improved compared to the control plants, especially for the doses of 5%, 10%, 20%, and 25% of compost. Similarly, the nutritional quality of mature fava bean seeds showed significantly higher quality characteristics in terms of protein, lipids, total sugars, and fibers compared to seeds grown in the control group, especially for seeds from plants fertilized with 10% compost. In this regard, it can be said that the optimal dose under conditions similar to those in this experiment for the germination of fava beans is 10% compost, which corresponds to 4.32 g/m^2. The results of this study are in agreement with those of other research, which have also shown that the use of olive pomace compost as a soil amendment can be a sustainable and cost-effective alternative to polluting chemical fertilizers. Therefore, this practice can contribute to promoting more environmentally friendly and sustainable agriculture.

REFERENCES

A.O.A.C. 1990. *Official Methods of Analysis*. 15th Edition, Association of Official Analytical Chemist, Washington DC.

Alaoui, Si Bennasseur, 2005. Référentiel pour la Conduite Technique de la fève (Vicia faba). 9–101.

Alburquerque, J.A., de la Fuente, C., Bernal, M.P., 2011. Improvement of soil quality after "alperujo" compost application to two contaminated soils characterised by differing heavy metal solubility. *J. Environ. Manage.* 92, 733–741.

Alburquerque, J.A., Gonzálvez, J., García, D., Cegarra, J., 2006. Composting of a solid olive-mill by-product and the potential of the resulting compost for cultivating pepper under commercial conditions. *Waste Manag.* 6, 620–626.

Alburquerque, J.A., Gonzalvez, J., Garcìa, D., Cegarra, J., 2007. Effects of a compost made from the solid byproduct (alperujo) of the two-phase centrifugation system for olive oil extraction and cotton gin waste on growth and nutrient content of ryegrass (*Lolium perenne* L.). *Bioresour. Technol.* 98 (4), 940–945.

Ameziane, H., Nounah, A., Kabbour, M.R., Khamar, M. 2020a. Agronomic valuation of olive pomace obtained by different extraction systems. *Eco. Env. & Cons.* 26(1), 414–422.

Ameziane, H., Nounah, A., Khamar, M., Zouahri, A. 2020b. Composting olive pomace evolution of organic matter and compost quality. *Agron. Res.* 18(1), 517.

Ameziane, H., Nounah, A., Khammar, M., Cherkaoui, E., Kabbour, M.R. 2018. Agrochemical characterization of olive pomace obtained by different systems of extraction. *Der Pharma Chemica.* 10(12), 3340.

Baeta-Hall L, Sàágua MC, Bartolomeu ML, Anselmo AM, Rosa MF 2005. Biodegradation of olive oil husks in composting aerated piles. *Bioresour. Technol.* 96, 69–78.

Bové J.M. 1962. Quelques aspects anciens et modernes de la photosynthèse. *Fruits* 17(2), 55–74.

Bremner, J.M., Mulvaney, C. S. 1982. Nitrogen-total. In: *Methods of Soil Analysis, part 2*. A.L. Page, et al. (eds.). Agronomy Monog. 9 ASA and SSSA, Madison, WI, pp. 595–624.

Brunetti, G., Plaza, C., Senesi, N., 2005. Olive pomace amendment in Mediterranean conditions: Effect on soil and humic acid properties and wheat (*Triticum turgidum* L.). *J. Agric. Food Chem.* 53, 6730–6737.

Cao, Y., Gao, Y., Qi, Y., Li, J. 2018. Biochar-enhanced composts reduce the potential leaching of nutrients and heavy metals and suppress plant-parasitic nematodes in excessively fertilized cucumber soils. *Environ. Sci. Pollut. Res.* 25(8), 7589–7599.

Cayuela, M.L., Bernal, M.P., Roig, A., 2004. Composting olive mill waste and sheep manure for orchard use. *Compost. Sci. Util.* 12, 130–136.

Clemente, R., Walker, D.J., Pardo, T., Martínez-Fernández, D., Bernal, M.P., 2012. The use of a halophytic plant species and organic amendments for the remediation of a trace elements-contaminated soil under semi-arid conditions. *J. Hazard. Mater.* 223–224, 63–71.

Comoretto, L., Arfib, B., Chiron, S. 2007. Pesticides in the Rhône river delta (France): Basic data for a field-based exposure assessment. *Sci. Total Environ.* 380(1–3), 124–132. DOI: 10.1016/j.scitotenv.2006.11.046

Curaqueo, G., Schoebitz, M., Borie, F., Caravaca, F., Roldán, A., 2014. Inoculation with arbuscular mycorrhizal fungi and addition of composted olive-mill waste enhance plant establishment and soil properties in the regeneration of a heavy metal-polluted environment. *Environ. Sci. Pollut. Res.* 21, 7403–7412.

Del Buono, D., Said-Pullicino, D., Proietti, P., Nasini, L., Gigliotti, G. 2011. Utilization of olive husks as plant growing substrates: phytotoxicity and plant biochemical responses. *Compost Sci. Util.* 19, 52–60.

Dubois, M., Gilles, K.A., Hamilton, J.K., Rebers, P.A., Smith, F. (1956) Colorimetric method for determination of sugars and related substances. *Anal. Chem.* 28, 350–356. http://dx.doi.org/10.1021/ac60111a017

Filippi, M., Rocca, M.A., Colombo, B., et al. 2002. Functional magnetic resonance imaging correlates of fatigue in multiple sclerosis. *Neuroimage* 15, 559–567. http://doi.org/10.1006/nimg.2001.1011

García-Gomez, A., Bernal, M.P., Roig, A., 2002. Growth of ornamental plants in two composts prepared from agroindustrial wastes. *Bioresour. Technol.* 83, 81–87.

Gigliotti, G., Proietti, P., Said-Pullicino, D., Nasini, L., Pezzolla, D., Rosati, L., Porceddu, P.R. 2012. Co-composting of olive husks with high moisture contents: organic matter dynamics and compost quality. *Int. Biodeter. Biodegr.* 67, 8–14.

Innangi, M., Danise, T., d'Alessandro, F., Curcio, E., Fioretto, A., 2017. Dynamics of organic matter in leaf litter and topsoil within an Italian alder (alnuscordata (loisel.) desf.) ecosystem. *Forests* 8, 240. https://doi.org/10.3390/f8070240

ISO 16472:2006 Aliments des animaux – Détermination du contenu en fibre par traitement à l'amylase et au détergent neutre (aNDF).

Jake, Munroe. 2018. Ministère de l'Agriculture, de l'Alimentation et des Affaires rurales de l'Ontario. *Manuel sur la fertilité du sol.* Publication 611F, 3e édition. 2018, pp. 256.

Lanfer-Marquez, U.M., Barros, R.M.C., Sinnecker, P. 2005. Antioxidant activity of chlorophylls and their derivatives. *Food Res. Int.* 38(8–9), 885–891. DOI: 10.1016/j.foodres.2005.02.012

López-Piñeiro, A., Albarràn, A., Rato Nunes, J.M., Barreto, C., 2008. Short and medium term effects of two phase olive mill waste application on olive grove production and soil properties under semiarid Mediterranean conditions. *Bioresour. Technol.* 99, 7982–7987.

Martínez-Blanco, J., Antón, A., Rieradevall, J., Castellari, M., Muñoz, P., 2011.Comparing nutritional value and yield as functional units in the environmental assessment of horticultural production with organic or mineral fertilization. The case of Mediterranean cauliflower production. *Int. J. Life Cycle Assess.* 16 (1), 12–26. DOI: 10.1016/j.jclepro.2010.11.018

Martínez-Fernández, D., Walker, D.J., 2012. The effects of soil amendments on the growth of Atriplex halimus and Bituminaria bituminosa in heavy metal-contaminated soils. *Water Air Soil Pollut.* 223, 63–72.

Mekki, A., Dhouib, A., Sayadi, S., 2013. Review: effects of olive mill wastewater application on soil properties and plants growth. *Int. J. Recycl. Org. Waste Agric.* 2, 15–21.

Mérigout, P. 2006. Étude du métabolisme de la plante en réponse à l'apport de différents fertilisants et adjuvants culturaux: Influence des phytohormones sur le métabolisme azoté. PhD dissertation, Institut National Agronomique Paris-Grignon, France.

Nasini, L., Gigliotti, G., Balduccini, M.A., Federici, E., Cenci, G., Proietti, P., 2013. Effect of solid olive-mill waste amendment on soil fertility and olive (*Olea europaea* L.) tree activity. *Agric. Ecosyst. Environ.* 164, 292–297.

Nicholls, C., Altieri, M. 2014. Agroecology: Designing climate change resilient small farming systems in the developing world. Agroecology for food security and nutrition, Proceedings of the FAO International Symposium, Rome, 271–295.

Parcevaux, S., Catsky, J. 1970. Methods and techniques for measuring leaf surfaces. In: *Techniques for Studying the Physical Factors of the Biosphere*. Ed. INRA., Paris, 493–499.

Pardo, T., Clemente, R., Alvarenga, P., Bernal, M.P., 2014b. Efficiency of soil organic and inorganic amendments on the remediation of a contaminated mine soil: II. *Biological and ecotoxicological evaluation. Chemosphere* 107, 101–108.

Pardo, T., Martínez-Fernández, D., Clemente, R., Bernal, M.P., Walker, D.J., 2014d. The use of olive-mill waste compost to promote the plant vegetation cover in a trace element-contaminated soil. *Environ. Sci. Pollut. Res.* 21, 1029–1038.

Proietti, P., Federici, E., Fidati, L., Scargetta, S., Massaccesi, L., Nasini, L., Regni, L., Ricci, A., Cenci, G., Gigliotti, G. 2015.Effects of amendment with oil mill waste and its derived compost on soil chemical and microbiological characteristics and olive (*Olea europaea* L.) productivity. *Agric. Ecosyst. Environ.* 207, 51–60.

Regni, L., Gigliotti, G., Nasini, L. & Proietti, P. 2016. Reuse of olive mill waste as soil amendment. In: *Olive Mill Waste: Recent Advances for Sustainable Management*, C.M. Galanakis Ed.; Elsevier-Academic Press: Oxford, UK, pp. 97–117.

Regni, L., Nasini, L., Ilarioni, L., Brunori, A., Massaccesi, L., Agnelli, A. & Proietti, P. 2017. Long term amendment with fresh and composted solid olive mill waste on olive grove affects carbon sequestration by prunings, fruits and soil. *Front. Plant Sci.* 7, 20–42.

Secretariat of State at the Ministry of Energy, Mines, Water and Environment, in charge of Water and Environment. Quality Standards: Water for irrigation. 2007.

Tajada, M., Gonzalez, J.L., 2004. Effects of application of a by-product of the two-step olive oil mill process on maize yield. *Agron. J.* 96, 692–699.

Walker, D.J., Bernal, M.P., 2008. The effects of olive mill waste compost and poultry manure on the availability and plant uptake of nutrients in a highly saline soil. *Bioresour. Technol.* 99, 396–403.

CHAPTER 4

Modelling and Simulation of the Behavior of a Photovoltaic System under Different Shading Rates

Mohammed Benchrifa

University Mohammed V in Rabat, Morocco
Ibn Tofaïl University—Kenitra-University Campus, Kenitra, Morocco

Jamal Mabrouki, and Fatine Drhimer

University Mohammed V in Rabat, Morocco

4.1 INTRODUCTION

In recent years, the world demand for energy has grown significantly in all areas, including industry, agriculture, transport ....[1–3]. Unfortunately, most of this energy comes from fossil fuels, which have a very harmful impact on the environment. Therefore, in efforts to minimize the impact of pollution on the environment, renewable energy is the ideal solution. Consequently, several countries have made huge investments into the development of renewable energies [4, 5]. According to the definition, renewable energy refers to the ways of producing energy from sources or resources that are theoretically unlimited, and available without time limits, or which can be replenished more quickly than they are consumed.

Renewable energy is generally considered in opposition to fossil fuel energy, which has limited stocks and is non-renewable on a human timescale [6, 7]. This new type of energy includes wind energy (from the wind), tidal energy (from waves and tidal movements) and solar photovoltaic energy. This photovoltaic energy consists of the direct transformation of sunlight into electricity, using a semi-conductor material.

The photovoltaic effect was initially discovered by E. Becquerel in 1839 [8, 9], who discovered that certain materials would deliver a small amount of electricity when exposed to light. Later, in 1912, Albert Einstein explained the photoelectric phenomenon [10], but it

DOI: 10.1201/9781003436218-4

was not until the early 1950s that it was put into practice in a silicon photovoltaic (PV) cell, which had an efficiency level of 4.5% [11, 21].

Photovoltaics no longer plays a marginal role in the proportion of electricity production in many countries, and it use continues to grow very rapidly. However, this source of electricity tends to be highly variable and decentralized, which can cause problems when attempting to integrate this type of production into the grid. It is therefore becoming increasingly important to understand the behavior of photovoltaic modules under different conditions to ensure the reliability of production and facilitate its integration into the grid [12–15].

A photovoltaic system consists of several components: the photovoltaic modules in series-parallel; the inverter, which transforms the direct electrical energy from the modules into alternating electrical energy; a cabling system; and batteries or a transmission network [16–22]. In photovoltaic technology, it is clear that modules do not necessarily maintain their initial performance. Some modules may degrade or even fail when operated on-site for extended periods. Several studies have found that the reliability of PV systems is highly dependent on the material used to construct the PV panels, temperature, humidity and solar radiation. A PV system can have several faults, be they construction faults, or material and electrical faults caused by climatic conditions [23–25].

In this context, we can note that the most common defect in a PV generator is shading failure [26–31]. Shading failure is usually the result of a variety of causes, for example, bird droppings, dust, snow and obstacles such as vegetation and power cables... In fact, in this chapter we are particularly interested in the effect of shading on the performance of PV modules. In order to visualize this effect we performed several simulations of the behavior of a PV module by varying the shading rate from 10% to 100%.

4.2 MATERIALS AND METHODS

4.2.1 Presentation of MATLAB/Simulink

Simulink is a functional diagramming environment for multi-domain simulation and model-based design. It supports system-level design and simulation, automatic code generation, and th continuous testing and verification of embedded systems [32].

4.2.2 Presentation of PVSyst

The PVSyst software is a simulation and sizing software for both stand-alone and on-grid photovoltaics systems. It was established by Geneva University (Switzerland) and designed by A. Mermoud. The PVSyst software has several inputs: average monthly solar fluxes, average monthly temperatures, energy requirements, choice of PV modules and their inclination, choice of batteries, charge controllers, inverters, input of the number of days of autonomy, the solar coverage rate and the investment cost (material purchase, system installation cost) The main results of the simulation are: the required field power. The main results of the simulation are: the required field strength, the storage capacity, the characteristics of the components under specific conditions and the cost per kilowatt-hour (kWh) [33].

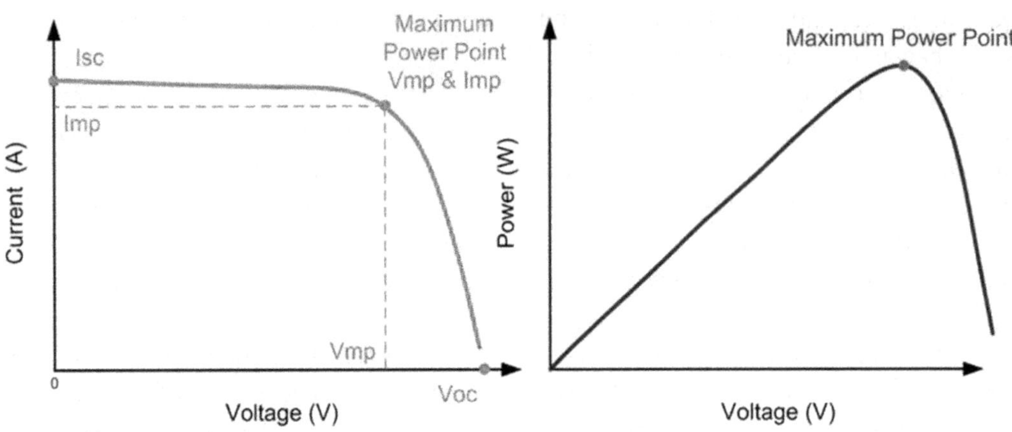

FIGURE 4.1 Current-voltage and power-voltage characteristic curves of a PV cell.

4.3 RESULTS AND DISCUSSION

4.3.1 Solar Cell Characteristics

Under a given illumination, a PV cell is characterized by a current-voltage curve (I-V), representing the set of electrical configurations that the cell can take. Three physical quantities define this curve. The first quantity is the open-circuit voltage Vco; this value represents the voltage generated by an illuminated cell that is not connected. The second quantity is the short-circuit current Icc; this value represents the current generated by an illuminated cell that is connected to itself. The third quantity is the maximum power point (PMPP) obtained for an optimal voltage and current: Vopt, Iopt (sometimes also called Vmpp, Impp).

We can also draw a characteristic curve representing the power as a function of the voltage. In this case, the maximum power reached for the voltage Vmp can be clearly observed in the (Figure 4.1).

a. **Shading effect on photovoltaic panels**

   Shade is the first enemy of a photovoltaic installation. A shaded cell will limit the generated power. It is therefore imperative to choose a location that is as free as possible from shadows caused by the environment.

   When a part of a photovoltaic unit is shaded, this under-irradiated part of the module can become inversely polarized. This means that the under-irradiated part behaves not as an electrical generator but as a receiver (resistance). The under-irradiated part will therefore behave as a receiver by dissipating a certain amount of power as heat, which will cause a rise of temperature in the under-irradiated area. This is the reverse self-polarization effect. This local heating can lead to hot spots which can damage the affected area and permanently degrade the performance of the photovoltaic module.

There are two types of shading: full shading and partial shading. Full shading prevents all radiation (direct and indirect) from reaching any part of the PV cell (e.g. bird droppings, tree branches on the panel, a cover). By contrast, partial shading prevents only direct radiation from reaching a part of the photovoltaic cell (e.g. a cloud).

Often the cells of a photovoltaic module are connected in series. Thus, the weakest cell will determine and limit the power of the other cells. Shading half of one cell or half of a cell row will decrease the power in proportion to the percentage of a cell's shaded area, in this case by 50%. The total shading of a cell row can reduce the power of the module to zero.

There are also three different shading configurations. The first configuration is single-cell shading, in which the procedure adopted ensures that only one solar cell of the module is shaded. The hidden part of the cell represents the shading rate. The second configuration is vertical shading (for N cells); in this case the shading is parallel to each shaded cell. The third configuration is perpendicular or horizontal shading (for N cells); in this pattern the procedure consists of shading the cells of the module with a single cell surface equivalent cover, so that the shading pattern is perpendicular to all cells, thus ensuring the same shading rate.

b. **Shading influence of on the optimal PMPP power simulation.**
   In order to illustrate the influence of shading on the behavior of a photovoltaic device, we will choose a model of 36 silicon-polycrystalline cells with a peak power of 60 W. We fix the cell temperature at 30°C. We assume that a cell is shaded by a percentage p. We determine the maximum power PMPP by changing the shading percentage by a step of 10%. The graphs we obtained are the following (Figure 4.2).

For each shading rate (Figure 4.2) there is a corresponding maximum power, after having determined these maximum powers for shading rates between 0% and 100%. The curve representing this variation is then drawn.

The (Figure 4.3) shows that in the absence of shading the PMPP is equal to the peak power of the module used (60 W). While this PMPP decreases as function of shading rate, it tends towards 0 for shading rate of 100%.

The general shape of the I(V) and P(V) characteristics of the photovoltaic panel shows significant deformations with increasing shading rate, which implies their non-uniformity. Indeed, the reduction of the short-circuit current of the shaded cell is evident and the generated current has decreased with the increase of shading rate [34–38]. There is another effect due to the shading, that of the displacement of the maximum power point to lower voltage values [39–42]. An increase in the shading rate causes a power reduction that is dissipated as heat. This effect must be taken into consideration, in the case of a system connected to a maximum power point tracking inverter, since it can cause a bad exploitation of the photovoltaic system [43–45].

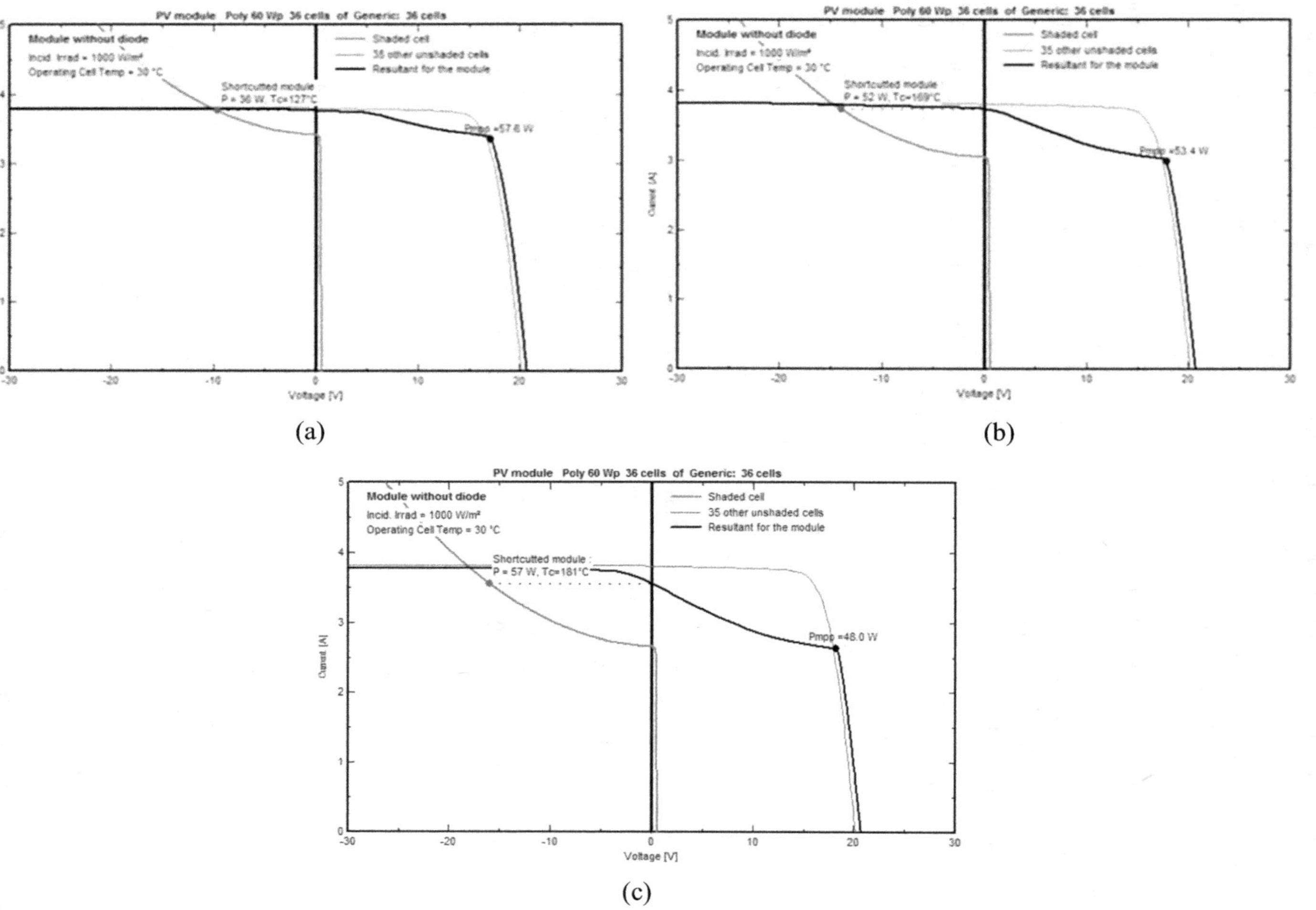

FIGURE 4.2 The variation of current as a function of voltage with an illuminance of 1000 W/m^2 and a shading rate of (a) 10%, (b) 20%, (c) 30%.

(Continued)

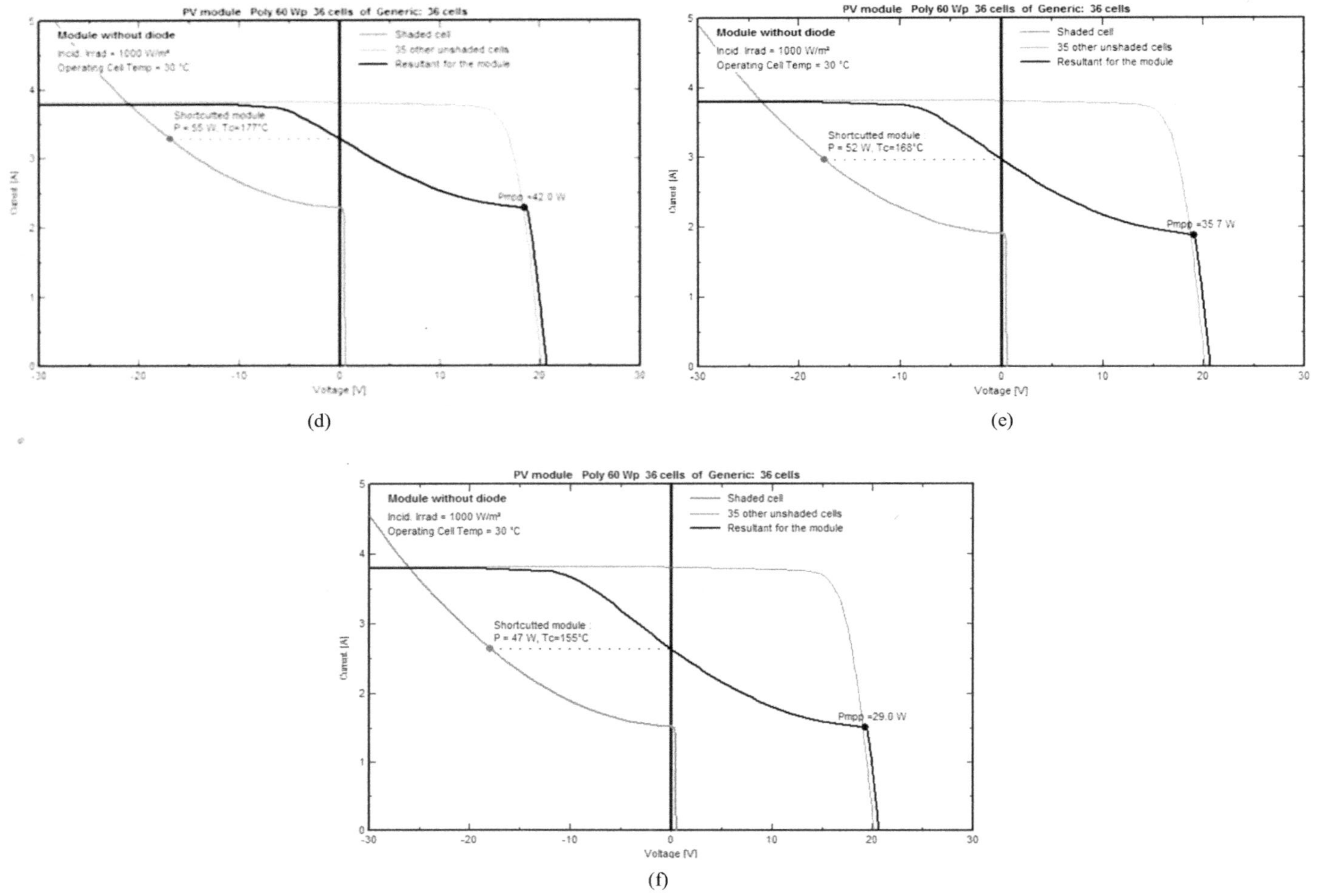

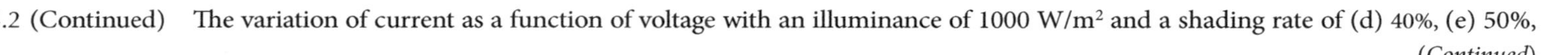

FIGURE 4.2 (Continued) The variation of current as a function of voltage with an illuminance of 1000 W/m² and a shading rate of (d) 40%, (e) 50%, (f) 60%.

(Continued)

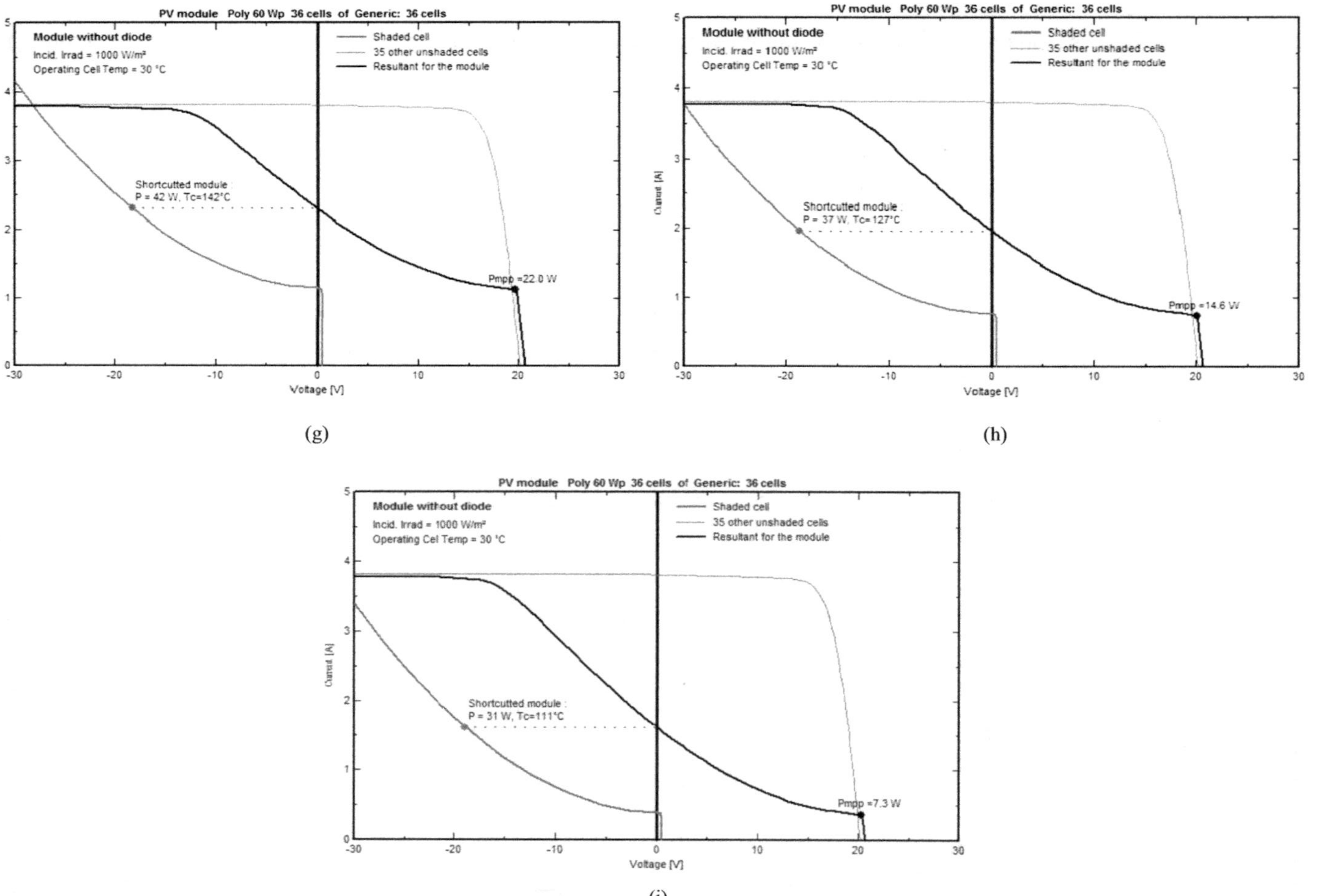

FIGURE 4.2 (Continued) The variation of current as a function of voltage with an illuminance of 1000 W/m^2 and a shading rate of (g) 70%, (h) 80 %, (i) 90%.

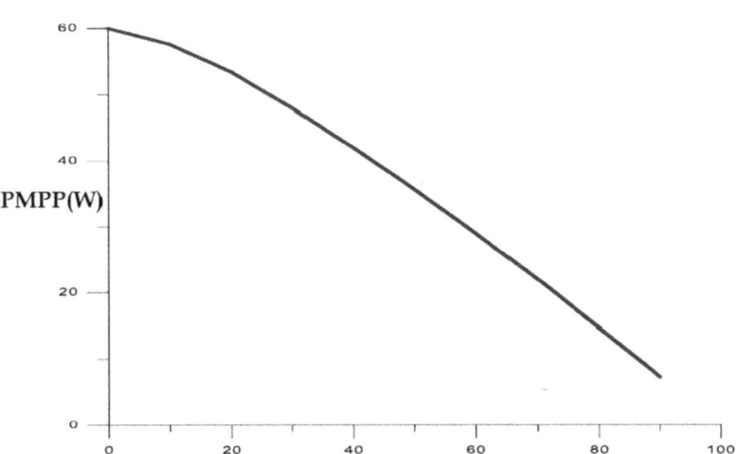

FIGURE 4.3 Maximum power (PMPP) variation.

4.4 CONCLUSION

The performance of a PV array is strongly influenced by climatic conditions, particularly solar irradiation. In this study, we used Matlab Simulink to simulate the operation of PV modules under different shading conditions. The main advantage of this model lies in its simplicity and ease of implementation, based on the technical specifications provided by the manufacturer. The result of this study showed that the generated current decreases as the shading rate increases, and that the PMPP drops from 60 W for 0% shading to 10 W for 90% shading. From this we can conclude that the shaded cell will limit the power generated by the whole panel and thus reduce its efficiency. Therefore, by this paper we were able to quantify and visualize the effect of shading on a photovoltaic panel under standard conditions of temperature and radiation

REFERENCES

[1] Shahzad, Umair The need for renewable energy sources. *Energy*, 2012, vol. 2, pp. 16–18.

[2] Prieto, Pedro A., Hall, Charles AS, and Melgar, Rigoberto. *Spain's photovoltaic revolution: The energy return on investment*. New York: Springer, 2013.

[3] Franzitta, Vincenzo, Curto, Domenico, Rao, Davide, et al. Hydrogen production from sea wave for alternative energy vehicles for public transport in Trapani (Italy). *Energies*, 2016, vol. 9, no 10, p. 850.

[4] Mekhilef, Saidur, Saidur, Rahman, and Safari, Azadeh. A review on solar energy use in industries. *Renewable and Sustainable Energy Reviews*, 2011, vol. 15, no 4, pp. 1777–1790.

[5] Kamalapur, G. D. and Udaykumar, R. Y. Rural electrification in India and feasibility of photovoltaic solar home systems. *International Journal of Electrical Power & Energy Systems*, 2011, vol. 33, no 3, pp. 594–599.

[6] Reich, Nils H., Mueller, Bjoern, Armbruster, Alfons, et al. Performance ratio revisited: Is PR > 90% realistic. *Progress in Photovoltaics: Research and Applications*, 2012, vol. 20, no 6, pp. 717–726.

[7] Triki-Lahiani, Asma, Abdelghani, Afef Bennani-Ben, and Slama-Belkhodja, Ilhem. Fault detection and monitoring systems for photovoltaic installations: A review. *Renewable and Sustainable Energy Reviews*, 2018, vol. 82, pp. 2680–2692.

[8] Park, Nam-Gyu, Miyasaka, T., and Grätzel, M. *Organic-inorganic halide perovskite photovoltaics*. Cham, Switzerland: Springer, 2016.

[9] Prince, Morton B. *Early work on photovoltaic devices at the Bell telephone laboratories. Power for the world. The emergence of electricity from the sun*. Singapore: Pan Standorf Pub, 2011, pp. 497–498.

[10] Mulvaney, Dustin. *Solar power: Innovation, sustainability, and environmental justice*. University of California Press, 2019.

[11] Shengule, Rohit D. Progress of solar renewable energy: Research and achievement in particular reference to the world and India. In: AIP Conference Proceedings. AIP Publishing LLC, 2021. p. 020208.

[12] Millikan, Robert Andrews. *Radio's Past and Future*. University of Chicago Press, 1931.

[13] Kazmerski, Lawrence L. Innovation in Solar Technology: Toward a 100% Renewable Electricity Future. In: *Sustainable energy development and innovation*. Cham: Springer, 2022. pp. 3–12.

[14] Grätzel, Michael. Photoelectrochemical cells. In: *Materials for sustainable energy: A collection of peer-reviewed research and review articles from nature publishing group*, 2011. pp. 26–32.

[15] Bagher, Askari Mohammad, Vahid, Mirzaei Mahmoud Abadi, and Mohsen, Mirhabibi. Types of solar cells and application. *American Journal of optics and Photonics*, 2015, vol. 3, no 5, pp. 94–113.

[16] Butler, Keith T., Frost, Jarvist M., and Walsh, Aron. Ferroelectric materials for solar energy conversion: Photoferroics revisited. *Energy & Environmental Science*, 2015, vol. 8, no 3, pp. 838–848.

[17] Degiorgio, Vittorio and Cristiani, Ilaria. Semiconductor devices. In: *Photonics*, edited by Neil Ashby, William Brantley, Matthew Deady, Michael Fowler, Morten Hjorth-Jensen, Michael Inglis, Barry Luokkala. Cham: Springer, 2016. pp. 145–169.

[18] Munzer, K. Adolf, Holdermann, Konstantin T., Schlosser, Reinhold E., et al. Thin monocrystalline silicon solar cells. *IEEE Transactions on Electron Devices*, 1999, vol. 46, no 10, pp. 2055–2061.

[19] Bergmann, R. B., Berge, C., Rinke, T. J., et al. Advances in monocrystalline Si thin film solar cells by layer transfer. *Solar Energy Materials and Solar Cells*, 2002, vol. 74, no 1–4, pp. 213–218.

[20] Chen, Kejun, Bothwell, Alexandra, Guthrey, Harvey, et al. Measurement of poly-Si film thickness on textured surfaces by X-ray diffraction in poly-Si/SiOx passivating contacts for monocrystalline Si solar cells. *Solar Energy Materials and Solar Cells*, 2022, vol. 236, p. 111510.

[21] Abou Jieb, Yaman and Hossain, Ekram. *Photovoltaic systems: Fundamentals and applications*. Switzerland: Springer Nature, 2022.

[22] Lee, Chung Geun, Shin, Woo Gyun, Lim, Jong Rok, et al. Analysis of electrical and thermal characteristics of PV array under mismatching conditions caused by partial shading and short circuit failure of bypass diodes. *Energy*, 2021, vol. 218, p. 119480.

[23] Arekar, Kaustubh, Puranik, Vishal E., and Gupta, Rajesh. Performance analysis of PID affected crystalline silicon PV module under partial shading condition. In: 2021 IEEE 48th Photovoltaic Specialists Conference (PVSC). IEEE, 2021. p. 2415–2420.

[24] Chaibi, Y., Malvoni, M., Chouder, A., et al. Simple and efficient approach to detect and diagnose electrical faults and partial shading in photovoltaic systems. *Energy Conversion and Management*, 2019, vol. 196, pp. 330–343.

[25] Hysa, Azem. Modeling and simulation of the photovoltaic cells for different values of physical and environmental parameters. *Emerging Science Journal*, 2019, vol. 3, no 6, pp. 395–406.

[26] Rasheed, Mohammed Siham and Shihab, Suha. Modelling and parameter extraction of PV cell using single-diode model. *Advanced Energy Conversion Materials*, 2020, pp. 96–104.

[27] Rasheed, Mohammed, Shihab, Suha, and Rashid, Taha. The Single Diode Model for PV Characteristics Using Electrical Circuit. *Journal of Al-Qadisiyah for Computer Science and Mathematics*, 2021, vol. 13, no 1, pp. 131–138.
[28] Chandel, Tarana Afrin, Yasin, Mohd Yusuf, and Mallick, Md Arifuddin. Modeling and simulation of photovoltaic cell using single diode solar cell and double diode solar cell model. *Int. J. Innovative Technol. Explor. Eng.(IJITEE)*, 2019, vol. 8, no 10.
[29] Peter, A. Gbadega and Saha, A. K. Electrical characteristics improvement of photovoltaic modules using two-diode model and its application under mismatch conditions. In: 2019 Southern African Universities Power Engineering Conference/Robotics and Mechatronics/Pattern Recognition Association of South Africa (SAUPEC/RobMech/PRASA). IEEE, 2019. p. 328–333.
[30] Prakash, Saripalli Bhanu, Singh, Gagan, and Singh, Sonika. Modelling and Performance Analysis of Simplified Two-diode Model of Photovoltaic cells. *Frontiers in Physics*, 2021, vol. 9, p. 236.
[31] H.-L. Tsai, Tu, C.-S., and Su, Y.-J.. Development of generalized photovoltaic model using MATLAB/SIMULINK. *Proceedings of the world congress on Engineering and computer science*, 2008, vol. 2008, pp. 1–6.
[32] Sera, D. Real-time modelling, diagnostics and optimised MPPT for residential PV systems. Doctoral thesis, Aalborg University, 2009.
[33] Mabrouki, J., Azrour, M., Dhiba, D., Farhaoui, Y., and El Hajjaji, S. IoT-based data logger for weather monitoring using Arduino-based wireless sensor networks with remote graphical application and alerts. *Big Data Mining and Analytics*, 2021, vol. 4, no. 1, pp. 25–32.
[34] Mabrouki, J., Azrour, M., Fattah, G., Dhiba, D., and El Hajjaji, S. Intelligent monitoring system for biogas detection based on the Internet of Things: Mohammedia, Morocco city landfill case. *Big Data Mining and Analytics*, 2021, vol. 4, no. 1, pp. 10–17.
[35] Mabrouki, J., Fattah, G., Al-Jadabi, N., Abrouki, Y., Dhiba, D., Azrour, M., and Hajjaji, S. E. Study, simulation and modulation of solar thermal domestic hot water production systems. *Modeling Earth Systems and Environment*, 2022, vol. 8, no. 2, pp. 2853–2862.
[36] Mabrouki, J., Azrour, M., Farhaoui, Y., and El Hajjaji, S. Intelligent system for monitoring and detecting water quality. In International Conference on Big Data and Networks Technologies. Springer, Cham, 2019, April, pp. 172–182.
[37] Mabrouki, J., Azoulay, K., Elfanssi, S., Bouhachlaf, L., Mousli, F., Azrour, M., and Hajjaji, S. E. Smart system for monitoring and controlling of agricultural production by the IoT. In IoT and Smart Devices for Sustainable Environment. Springer, Cham, 2022, pp. 103–115.
[38] Mabrouki, J., Azrour, M., and Hajjaji, S. E. Use of internet of things for monitoring and evaluating water's quality: A comparative study. *International Journal of Cloud Computing*, 2021, vol. 10, no. 5–6, pp. 633–644.
[39] Mabrouki, J., Benbouzid, M., Dhiba, D., and El Hajjaji, S. Simulation of wastewater treatment processes with Bioreactor Membrane Reactor (MBR) treatment versus conventional the adsorbent layer-based filtration system (LAFS). *International Journal of Environmental Analytical Chemistry*, 2020, pp. 1–11.
[40] Benchrifa, M., Mabrouki, J., Elouardi, M., Azrour, M., and Tadili, R. Detailed study of dimensioning and simulating a grid-connected PV power station and analysis of its environmental and economic effect, case study. *Modeling Earth Systems and Environment*, 2022, vol. 9, pp. 1–9.
[41] Benchrifa, M., and Mabrouki, J.. Simulation, sizing, economic evaluation and environmental impact assessment of a photovoltaic power plant for the electrification of an establishment. *Advances in Building Energy Research*, 2022, pp. 1–18.
[42] Mabrouki, J., Azrour, M., Boubekraoui, A., and El Hajjaji, S. Intelligent system for the protection of people. In: *Intelligent systems in big data, semantic web and machine learning*, edited by Noreddine Gherabi and Janusz Kacprzyk. Cham: Springer, 2021, pp. 157–165.

[43] Mabrouki, J., Azrour, M., Dhiba, D., and Hajjaji, S. E.. High-fidelity intelligence ventilator to help infect with COVID-19 based on artificial intelligence. In: *Intelligent data analysis for COVID-19 pandemic*, edited by M. Niranjanamurthy, Siddhartha Bhattacharyya and Neeraj Kumar. Singapore: Springer, 2021, pp. 83–93.
[44] Benchrifa, M., Essalhi, H., Tadili, R., and Nfaoui, H. Estimation of daily direct solar radiation for Rabat. In: *Sustainable Energy Development and Innovation*, edited by Ali Sayigh. Cham: Springer, 2022, pp. 629–634.
[45] Benchrifa, M., Essalhi, H., Tadili, R., Bargach, M. N., and Mechaqrane, A. (2019). Development of a daily databank of solar radiation components for Moroccan territory. *International Journal of Photoenergy, vol.* 2019.

CHAPTER 5

Geochemical Study of the Angads Groundwater (North-East Morocco)

Application of the Piper and Schoeller-Berkaloff Diagram

Latifa Taoufiq, Ilias Kacimi, Mohamed Saadi, and Jamal Mabrouki
University Mohammed V in Rabat, Morocco

Nordine Nouayti
Abdelmalek Essaadi University, Al-Hoceima, Morocco

Nadia Kassou and Tarik Bouramtane
University Mohammed V in Rabat, Morocco

Karima El-Mouhdi
Higher Institute of Nursing Professions and Healthcare Technics, Meknes, Morocco

5.1 INTRODUCTION

All over the world, the use of water resources, mainly groundwater, is increasing, due to socio-economic and demographic development, as well as the degradation in the quality of surface water.

In Morocco, water availability is very limited and is likely to decrease in the long term due to global warming (1). Indeed, the forecast estimates made by the World Bank in 1994 sounded the alarm. Furthermore, renewable water resources per capita are likely to decrease by half, from 800 m^3 in 1990 to 400 m^3 in 2020, thus classifying Morocco in the

DOI: 10.1201/9781003436218-5

category of countries in a situation of chronic water stress (2). Indeed, the composition of groundwater reflects interactions with the vegetation cover, the characteristics of the soil through which it flows, the inputs from the atmosphere, the presence of industrial or agricultural activities, and the lithology of the geological formations that host the aquifers (3, 4). Moreover, groundwater is traditionally the water resource for human consumption and irrigation, as it is less polluting than surface water (5, 6). The degradation of groundwater quality is caused by various sources of pollution, such as fertilizers, pesticides, untreated water discharges, etc., as well as the geological formations that make up the groundwater. Hence the need for physicochemical and bacteriological analyses, which constitute a reliable source of information to characterize the impact of human, agricultural, and industrial activities and of the different landscape units (7), as well as the origin of mineralization.

The objective of this work is to make a geochemical study of the groundwater of the Angads aquifer based on the Piper diagram and the Schoeller-Berkaloff diagram. The aim is to know the dominant facies in our study area, in order to understand the current state of mineralization of this aquifer to protect it from any type of contamination.

5.2 MATERIALS AND METHODS

5.2.1 Study Site

The study site is represented by the Angads aquifer located at the eastern end of the Taourirt Oujda corridor (Figure 5.1). This aquifer covers an area of 460 km^2 (8), and is bounded to the south by Jbel Hamra, to the west by Jbels Megrez and Tarraza, to the north by the Beni Snassen chain (9), and to the east by the Mernia plain (10).

5.2.2 Conduct of the Study

The study was conducted in 2020. Nineteen sampling points were taken throughout the study area, distributed as follows: (a) two boreholes, (b) one well located near and the other downstream of the Oujda public dump, seventeen wells including six wells upstream and downstream of the wastewater treatment plant, for the other wells are distributed throughout the study area. The choice of these points is explained by the fact that they are the most vulnerable to pollution and the most used in various fields, whether domestic, industrial, or agricultural.

The geochemical study of groundwater has been represented by two diagrams, the first of which is Piper, which represents the chemical facies from a graphic distribution of major cations (Ca^{2+}, Mg^{2+}, Na^+ and K^+) and major anions (HCO_3^-, Cl^-, SO_4^{2-} and NO_3^-) (11). This diagram is made up of two triangles representing the anionic and cationic facies and a rhombus synthesizing the overall facies (12). The second diagram is the Schoeller-Berkaloff diagram which characterizes the chemical facies of several glasses of water (13). The cartographic representation of the data was carried out using a Geographic Information System (GIS). So far, no study has been carried out with a geochemical study. This is an important study to know the state of mineralization of groundwater, which is an added value for these resources.

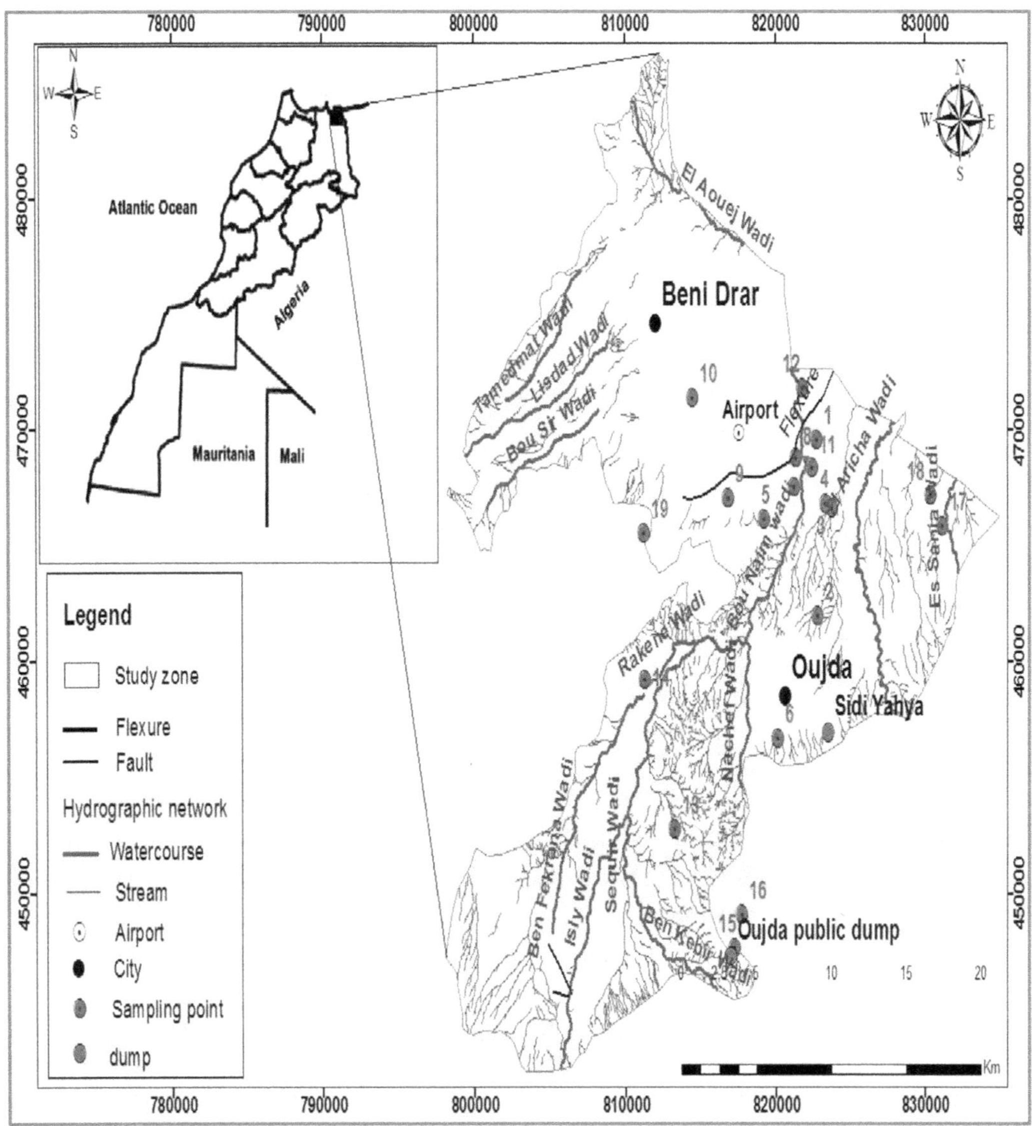

FIGURE 5.1 Map of the location and hydrological network of the aquifer (14).

5.3 RESULTS

5.3.1 Physicochemical and Bacteriological Analyses

The parametric results of the physicochemical analyses of the groundwater from nineteen sampling points of the Angads aquifer (Table 5.1) reveal that the Cond varies from 10,355 $\mu S.cm^{-1}$ to 986 $\mu S.cm^{-1}$, with an average of 2463 $\mu S.cm^{-1}$ and a standard deviation of 2172. The pH varies from 6.95 to 7.55 with a mean of 7.20 and a standard deviation of 0.20. For cations such as Na^+ the values vary from 44 mg/l to 1227 mg/l, with a mean of 215 mg/l and a standard deviation of 265.6, K^+ values vary between 0.72 mg/l and 15.30 mg/l with a mean of 6.38 mg/l and a standard deviation of 146.62. Then, for the anions such as NO_3^-,

TABLE 5.1 Physicochemical and Bacteriological Data of Angads Groundwater in 2020 (14)

Sample Points	Parameters	Units	Mean	Maximum	Minimum	Standard Deviation
1	pH	—	7.19	7.55	6.95	0.19
2	T°	°C	21.39	23.6	19.7	1.18
3	Cond	(μS/cm)	2463	10,355	986	2171.78
4	IP	(mg/l)	1.94	9.33	0.65	3.67
5	NH_4^+	(mg/l)	0.16	0.67	0.02	0.25
6	Na^+	(mg/l)	215.45	1226.5	43.7	265.58
7	K^+	(mg/l)	6.37	15.3	0.71	146.62
8	Ca_2^+	(mg/l)	162.88	601	72	146.62
9	Mg^{2+}	(mg/l)	152.21	693	54.67	154.91
10	Fe^{2+}	(mg/l)	1.41	17.7	0.06	4.23
11	Mn^{2+}	(mg/l)	0.27	0.46	0.07	0.27
12	Cl^-	(mg/l)	607.23	3597	172	788.87
13	NO_2^-	(mg/l)	16.23	176	0.011	52.99
14	NO_3^-	(mg/l)	50.62	134	5.38	39.8
15	SO_4^{2-}	(mg/l)	219.42	1395	49	323.46
16	HCO_3^-	(mg/l)	383.98	811.3	250.1	138.22
17	CT	(UFC/100 ml)	4668.52	36000	0	11,317.84
18	CF	(UFC/100 ml)	557.89	4800	0	1399.31
19	SF	(UFC/100 ml)	1130.63	9000	0	2505.68

the values vary between 5.4 mg/l and 134 mg/l with a mean of 50.6 mg/l and a standard deviation of 39.80, and for Cl^-, the values vary 172 mg/l and 3597 with a mean of 607 mg/l and a standard deviation of 788.9.

5.3.2 Hydrochemical Analysis

The chemical analyses are illustrated on the Piper triangle diagram (Figure 5.2) and the Schoeller-Berkaloff diagram. The results of the latter show that the water has identical profiles in the majority of the sampling points. With the exception of the four points 7, 8, 10 (Figure 5.3a), and 15 (Figure 5.3b) which present high values in some chemical elements, namely point 7 in Na^+, K^+, Cl^-, HCO_3^-, CO_3^-, for point 8 in Mg^{2+}, Cl^-, for point 10 in Na^+, K^+, Cl^- and for point 15 in Mg^{2+}, Na^+, K^+, Cl^-. The Piper diagram results reveal the appearance of two types of hydrogeochemical facies. The most important of these is the chloride and sulphate calcic and magnesian facies found at sampling points 1, 2, 3, 4, 5, 6, 8, 9, 11, 12, 13, 14, 15, 16, 17, 18 and 19, occupying 89.47%, and the other is a chloride sodium and potassium facies at points 7 and 10 with a percentage of 10.52%.

5.4 DISCUSSION

The representation of the physicochemical analyses of the groundwater from sampling points distributed throughout the Angads aquifer using the Piper diagram revealed the existence of chloride and sulphate calcic and magnesian facies with a percentage of 89.47%. This shows infiltration and dissolution of the Triassic argillo-evaporite rocks, it generally

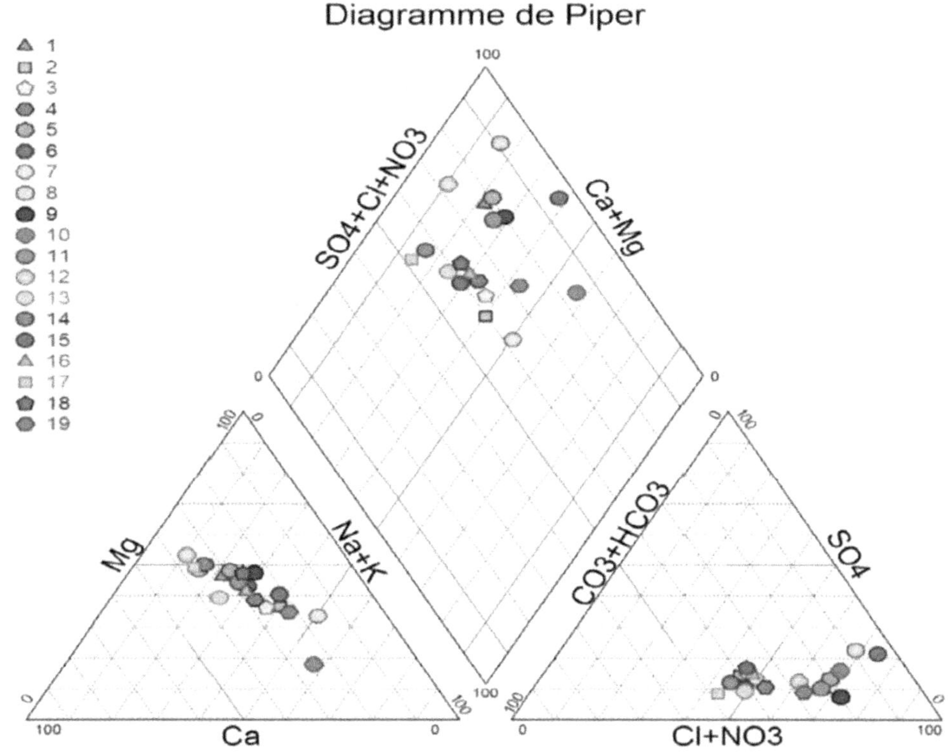

FIGURE 5.2 Piper diagram of Angads groundwater in September 2020 (14).

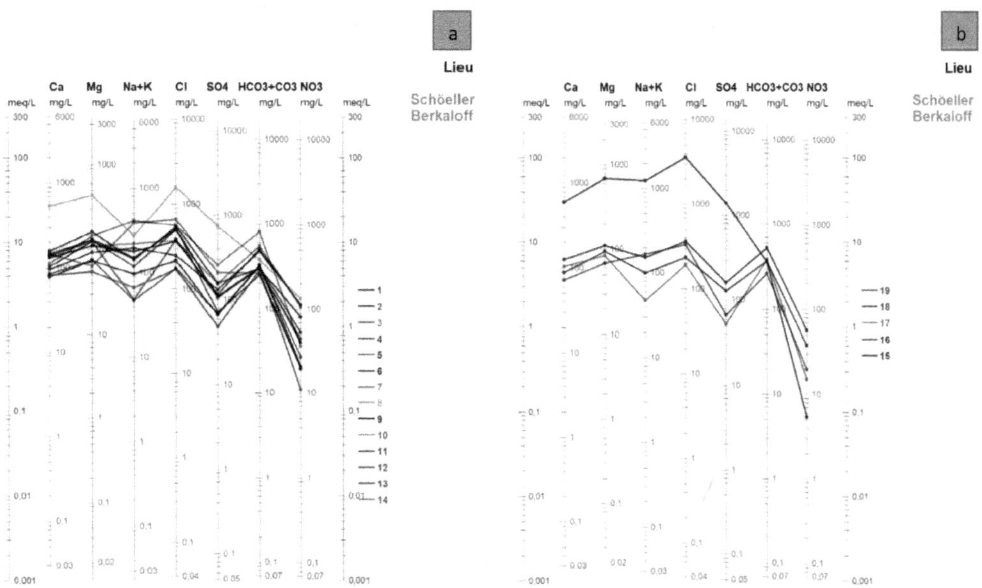

FIGURE 5.3 (a–b): Schoeller-Berkaloff diagram of Angads groundwater in September 2020 (14).

represents the mineralized waters. These results have been confirmed by previous studies such as Amrani's study, which presents hydrogeochemical and principal component analysis (PCA) in the explanation of the groundwater chemistry of the plio-quaternary aquifer between Timahdite and Almis Guigou (Middle Atlas, Morocco)(15) and Chaoui's study on the vulnerability to pollution of surface and groundwater in the Bouchegouf region (North-East Algeria) (16). The use of the Schoeller-Berkaloff diagram showed that the majority of the sampling points have the same patterns, with the exception of the four points 7, 8, 10, and 15. These points have high values of some chemical parameters, namely point 7 in Na^+, K^+, Cl^-, HCO_3^-, CO_3^-, point 8 in Mg^{2+}, Cl^-, and SO_4^{3-}, for point 10 in Na^+, K^+, Cl^- and for point 15 in Mg^{2+}, Na^+, K^+, Cl^-, and SO_4^{2-}. This explains why the groundwater in the Angads aquifer is highly mineralized. These results have been confirmed by previous studies such as the study by (15), and another study by Jilali, which presents the hydrochemistry and geothermometry of the thermal waters of northeast Morocco (17).

This study illustrates the hydrogeochemistry of the groundwater in our study area, in order to determine the mineralogical composition of this groundwater given the importance of this substance for the health of the population.

5.5 CONCLUSION

In this study, we analyzed the physicochemical parameters of the groundwater of the Angads aquifer. The representation of these analyses on the Piper diagram allows us to deduce the predominance of chloride and sulphate calcic and magnesian facies with a percentage of 89.47% at the level of sampling points 1, 2, 3, 4, 5, 6, 8, 9, 11, 12, 13, 14, 15, 16, 17, 18 and 19, and another sodium and potassium chloride facies at the two points with a percentage of 10.52% at two points, 7 and 10. The results of the Schoeller-Berkaloff diagram show that the majority of the sampling points have the same appearance, with the exception of the four points 7, 8, 10 and 15. These points have high values in some chemical parameters, namely, point 7 in Na^+, K^+, Cl^-, HCO_3^-, CO_3^-, point 8 in Mg^{2+}, Cl^-, and SO_4^{3-}, for point 10 in Na^+, K^+, Cl^-, and for point 15 in Mg^{2+}, Na^+, K^+, Cl^-, and SO_4^{2-}. This work confirms that the groundwater in our study area is mineral water, which enhances the value of this resource; in order to protect it, an effective and sustainable strategy against the different types of pollution must be considered.

ACKNOWLEDGMENTS

The authors would like to thank all the staff working at the Al-Hoceima-Tetouan Water and Environment Management, Laboratory and Geosciences Research Team, Water and Environment Laboratory of Rabat, who contributed to the realization of this study.

CONFLICTS OF INTEREST

There is no conflict of interest between the authors.

FUNDING

This study was not funded by any organization.

REFERENCES

1. Nouayti, N., Khattach, D., & Hilali, M. Assessment of physico-chemical quality of groundwater of the jurassic aquifers in high basin of ziz *Journal of Materials and Environmental Science* 2015;6(4):1068–1081.
2. Berdaï H. Synthesis of work carried out in Morocco on the nitrate pollution of groundwater. Studies Division, Experimentation, Testing and Standardization Department, Rabat. 1997.
3. Asano T, Cotruvo JA. Groundwater Recharge with Reclaimed Municipal Wastewater: Health and Regulatory Considerations. *Water Res* 2004;38(8):1941–1951.
4. Quiers M, Batiot-Guilhe C, Bicalho C, Perrette Y, Seidel J, Van-Exter S. Characterisation of Rapid Infiltration Flows and Vulnerability in a Karst Aquifer Using a Decomposed Fluorescence Signal of Dissolved Organic Matter. *Environ Earth Sci* 1 Jan. 2013;71:553–561.
5. Nouayti N, Khattach D, Hilali M, Brahimi A, Baki S. *Assessment of Metal Contamination in Jurassic Aquifers of Ziz High Basin* (Central High Atlas, Morocco). 2016;9.
6. Baki S, Hilali M, Kacimi I, Kassou N, Nouiyti N, Bahassi A. Assessment of Groundwater Intrinsic Vulnerability to Pollution in the Pre-Saharan Areas – The Case of the Tafilalet Plain (Southeast Morocco). *Procedia Earth Planet Sci* 2017;17:590–593.
7. Souley Moussa R, Malam Alma MM, Laouli MS, Natatou I, Habou I. Physico-Chemical Characterization of the Waters of the Continental Intercalaire/Hamadien and Continentalsiems Terminal Aquifers in the Zinder Region (Niger). *Int J Bio Chem Sci* 7 Jan 2019;12(5):2395.
8. Lahrach A, Zarhloule Y, Azhimi M, Boumashouli SM, Jabrane R, Dsouli K, et al. The Angads Groundwater (Eastern Morocco): Flow Modeling and Management Perspectives. *Geomaghreb – 2006* 2005;3:23–34.
9. Moulouya Hydraulic Basin Agency. *Inventory Study of Water Withdrawals from the Angads Aquifer* (Kingdom of Morocco Moulouya Hydraulic Basin Agency, Oujda, Morocco). 2017.
10. Es-Sousy M, Gharibi E, Ghalit M, Taupin JD, Ben Alaya M. Hydrochemical Quality of the Angads Plain Groundwater (Eastern Morocco). In: Conference of the Arabian Journal of Geosciences. Springer; 2018. p. 99–102.
11. Moussa RS, Alma MMM, Laouli MS, Natatou I, Habou I. Physico-Chemical Characterization of the Waters of the Continental Intercalaire/Hamadien and Continentalsiems Terminal Aquifers in the Zinder Region (Niger). *Int J Biol Chem Sci* 2018;12(5):2395–23411.
12. Bouchemal F, Achour S. Physico-Chemical Quality and Pollution Parameters of Groundwater in the Biskra Region. *LARHYSS J* 2015;22:197–212.
13. Dominique V. Studies of Physico-Chemical Data of the Waters of the Northern Sector of Piton des Neiges, Reunion Island, 2019.
14. Taoufiq L, Kacimi I, Saadi M, Nouayti N, Kassou N, Bouramtane T, et al. Assessment of Physicochemical and Bacteriological Parameters in the Angads Aquifer (Northeast Morocco): Application of Principal Component Analysis and Piper and Schoeller-Berkaloff Diagrams. *Applied and Environmental Soil Science*, 2023;2023, 14 pages, https://doi.org/10.1155/2023/2806854.
15. Amrani S, Hinaje S. Use of Hydro-Geochemical Analyzes and Principal Component Analyzes (PCA) in the Explanation of Groundwater Chemistry of the Plio-Quaternary Aquifer Between Timahdite and Almis Guigou (Middle Atlas, Morocco). *ScienceLib Editions Mersenne* 2014;6: 1–14.
16. Chaoui W, Bousnoubra H, Chaoui K. Study of the Vulnerability to Pollution of Surface and Underground Water in the Region of Bouchegouf (North-East Algeria). *Nature & Technology*, 2013;8:33B.
17. Jilali A, Chamrar A, El Haddar A. Hydrochemistry and Geothermometry of Thermal Water in Northeastern Morocco. *Geothermal Energy.* 2018;6:1–16.

CHAPTER 6

Remote Monitoring of an Inclined Solar Distiller

Souad Nasrdine

University Mohammed V in Rabat, Morocco

Mohammed Benchrifa

University Mohammed V in Rabat, Morocco
Ibn Tofaïl University—Kenitra-University Campus, Kenitra, Morocco

Najlaa Ben-Lhachemi, Jamal Mabrouki, and Miloudia Slaoui

University Mohammed V in Rabat, Morocco

6.1 INTRODUCTION

Water is essential for human life, it makes up 65% of the human body, and covers 70% of the Earth's surface. Lately, however, the availability of fresh water is decreasing from natural resources more and more due to pollution. Water, climate change and falling groundwater levels are observed all around the world. According to studies carried out, they have shown that the availability of water should decrease by between 10 and 30% in certain arid and semi-arid regions [1], which will lead us to look for effective and less energy-consuming solutions to alleviate water shortages. The most effective way to purify contaminated water on a small scale is solar-powered desalination, which uses a free, renewable energy source.

In the field of solar desalination [2], the solar still is frequently used because of its straightforward construction, but its performance is relatively low when compared to other technologies. Several study projects are currently in progress to improve the productivity of the simple solar still, which was invented in 1872, by the Swedish engineer Charles Wilson [3–5]. The solar still works according to the same principles as rain: evaporation and condensation. It uses the sun's rays to heat brackish water at the level of the basin, which evaporates, then condenses on the front part of the glass. The two main types of solar distillation systems currently available are passive systems and active systems [6]. Simple passive stills are the most practical solar distillation system because they are less expensive and require no other source of energy apart from the Sun to operate. There are several different types of such stills. However, although they may differ in terms of materials used, design and

DOI: 10.1201/9781003436218-6

functionality, they all follow the same basic principles. We find the single-effect still [7] and the double-effect solar still [8, 9]... However, active solar stills use an additional source of energy to speed up the evaporation process [10]. They may differ in terms of materials used, design, and functionality; they all follow the same basic principles.

The main objective of any solar distillation system is to have the maximum productivity. The performance of the still depends on several factors, such as meteorological parameters: ambient temperature, solar irradiation, glass temperature and other operating and construction parameters.

In this study, we used AI to optimize the efficiency of solar stills by integrating a solenoid valve to ensure the automatic filling of the tank, and we installed level sensors to automatically measure productivity and finally notifications of alerts are sent to be informed in real time of the problems detected by the AI.

6.2 LITERATURE REVIEW

Distillers have existed since antiquity in several conventional forms that have been developed over time by researchers to achieve a less expensive form using solar energy, but the major drawback of this type of distiller is the low production of distillate. However, many studies have continued to use various methods to increase the production of solar energy, including coupling the solar distiller with a preheating system such as solar collectors [11], using heat storage materials [12], nanomaterials [13], using a condenser to easily condense the steam inside the tank and collect the steam droplets [14]. Recent studies have been conducted using the Internet of Things and Artificial Intelligence to improve the performance of the solar distiller by developing a mathematical model to predict the performance of solar distillers [15].

6.3 PROPOSED SYSTEM

Our system is composed of a tilted solar still facing south with a corrugated zinc absorber to increase the absorption surface, an ordinary glass cover to act as a condenser, and two tanks: one dedicated for the feeding the distiller with salt water and the other placed at the ends to recover the distillate.

6.4 MATERIALS AND METHODS

6.4.1 The Operating Principle of a Solar Distiller

This process is used to heat the water by the energy of solar radiation, which passes through the transparent glass cover and falls on the corrugated zinc basin, which contains saline water. When water is heated, after a while it reaches a high temperature, which allows it to evaporate. The temperature difference between the surface of the water and the lower surface of the cover causes the condensation of steam on the lower part of the cover. This condensed water is collected in gutters.

6.4.2 Solenoid Valve

The solenoid valve is used to convert electrical energy into mechanical energy. It allows autonomous and remote control of the fluid flow in a system. The solenoid valve consists of two parts, as shown in Figure 6.1. The solenoid is made up of a coil which is used to operate the valve by passing an electric current through them to create an electromagnetic field and

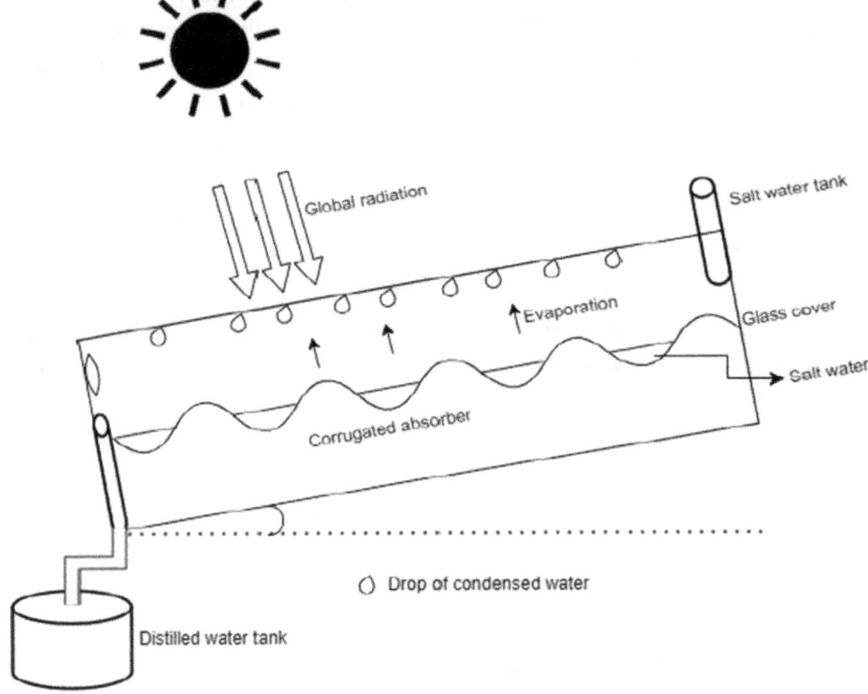

FIGURE 6.1 Diagram of the inclined solar distiller.

operate the valve, this means if it is connected to a controller it can be used in any way autonomous and remotely by a computer without the need to physically move to open and close the valves. This allows the system to operate much more efficiently and safely.

There are two types of valves: the normally open and normally closed type. In our case, we will use the normally closed valve (Figure 6.2).

FIGURE 6.2 Solenoid valve.

6.4.3 Level Sensor

A level sensor is an electronic device that measures the level of liquid in a tank or other container. Liquid level sensors can operate according to different measurement concepts, including pressure, electrical capacity, radiation, ultrasound, vibration, and electrical conductivity.

In this study we will rather be interested in the ultrasonic sensor which works by emitting a sound pulse which propagates in the air to the surface of the liquid and is reflected when it meets the surface of the liquid, the sensor detects the echo and measures the time it takes for the echo to return. The distance between the liquid surface and the sensor is calculated using the speed of sound in air. Figures 6.3 and 6.4 show respectively a sensor and its simplified operating diagram.

6.4.4 The Arduino Board

The Arduino board: In reality, it is an open-source platform called Arduino UNO, which is based on an 8-bit microcontroller and has additional parts that facilitate programming and connection to other electronic circuits (Figure 6.5).

The Arduino IDE, which allows you to edit, compile and transfer a program to its memory, can be used to program this microcontroller. Just use a USB/Serial cable to connect the Arduino motherboard to a computer [16].

FIGURE 6.3 Level sensor.

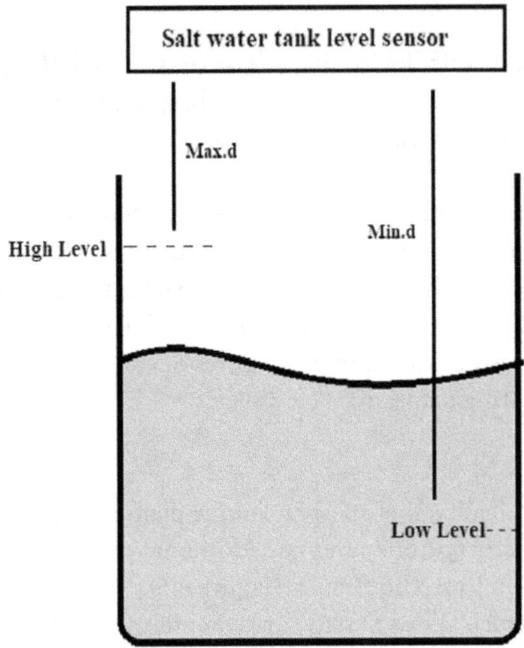

FIGURE 6.4 Operation of level sensor.

FIGURE 6.5 Arduino board.

6.5 RESULTS AND DISCUSSION

To better manage our system consisting of the solar still, we will automate the filling of the salt water tank by integrating a solenoid valve that we will control using an Arduino card.

6.5.1 Flowchart of the Solenoid Valve

The Figure 6.6 shows the flowchart that summarizes the operating program of the solenoid valve. In fact, when the water level in the tank is lower than the low level, the solenoid valve receives the signal to open and when the opposite occurs, the solenoid valve is always closed, the cycle repeats until the system stops.

Developing a monitoring system for the online control of the solar distiller is of great importance, because of improving the working conditions and productivity while monitoring the most important parameters. The realization of this assembly will allow us to automate a part of the distiller, and to control it remotely with the help of the Arduino board.

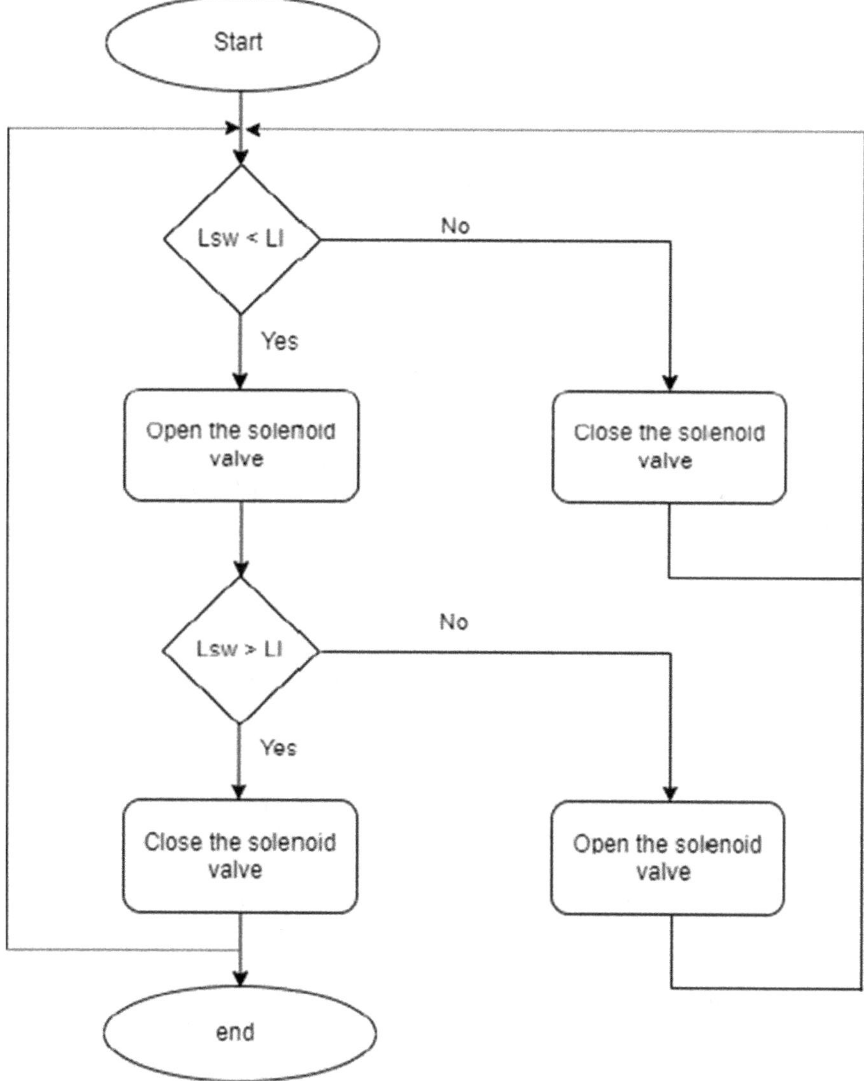

FIGURE 6.6 Flowchart of the solenoid valve. (Lsw: Salt water level; Ll: Low level, the low level in this case is determined from 1L.)

Thereafter we can also automate other parameters of the distiller, namely the temperature of the distiller, the solar radiation, the flow of distilled water, in order to arrive at a final autonomous system.

6.6 CONCLUSION

Human existence requires the consumption of energy and water. Researchers are working to identify strategies to meet the growing demand for water and electricity without endangering the environment. Among the alternatives found is solar desalination technology, which uses solar energy as a power source and is an efficient and environmentally friendly option. As part of this work, a distiller is already made with a zinc absorber; we have proposed a flowchart for a solenoid valve to be controlled with an Arduino card in order to automate the solar distiller, to increase the production of distilled water.

REFERENCES

[1] Z. W. Kundzewicz, L. J. Mata, N. W. Arnell, P. Döll, B. Jimenez, K. Miller, T. Oki, Z. Şen & I. Shiklomanov The implications of projected climate change for freshwater resources and their management. *Hydrological Sciences Journal* 2008;53.1: 3–10. DOI: 10.1623/hysj.53.1.3

[2] A. Y. Maalej Solar still performance. *Desalination* 1991;82:197–205.

[3] L. Cherrared Amélioration du Rendement d'un Distillateur Solaire à Effet de Serre. *Revue des Energies Renouvelables: Valorisation* 1999;121–124.

[4] N. Retiel, F. Abdessemed and M. Bettahar Etude expérimentale d'un distillateur solaire plan amélioré. *Revue des Energies Renouvelables* 2008;11(4):635–642.

[5] A. Chaker, G. Menguy Efficacité interne d'un distillateur solaire sphérique. energ.ren, 2001.

[6] M. Farid, F. Hamad. Performance of a single basin solar still, *Renewable Energy* 1993;3(4):75–83.

[7] T. Rajaseenivasan, K. Kalidasa Murugavel, T. Elango, R. Samuel Hansen. A review of different methods to enhance the productivity of the multi-effect solar still. *Renewable and Sustainable Energy Reviews* January 2013;17:248–259.

[8] R. V. Dunkle. Solar water distillation: The roof of type still and multiple effect diffusion still, International developments in heat transfer, A.S.M.E., Proc. International Heat Transfer, Part V, University of Colorado, 1961, p. 895.

[9] S. Kumar, G. N. Tiwari, H. N. Singh. Annual performance of an active solar distillation system. *Desalination* 2000;27:79–88.

[10] M. Patel, C. Patel, H. Panchal. Past, present and Future of Active solar distillation system: A comprehensive review. *International Journal of Ambient Energy* October 2019;43(2):1–23. DOI:10.1080/01430750.2019.1684996.

[11] A. Kaabi, O. Halloufi Etude de la performance d'un distillateur solaire par un système de pré-chauffage solaire de l'eau saumatre. 2017.

[12] N. Calvet, A. Meffre, R. Olivès, E. Guillot, X. Py, C. Bessada, P. Echegut. Matériaux de stockage thermique par chaleur sensible pour centrales électro-solaires testé sous flux solaire concentré. 2010.

[13] I. Alatawi, A. Khaliq, A. Mohamed, A. Heniegal, B. Abdelaziz, M. Elashmawy. Tubular solar stills: Recent developments and future. *Solar Energy Materials and Solar Cells* 2022;242:111785.

[14] H. Panchal, D. Mevada, K. K. Sadasivuni Recent advancements in condensers to enhance the performance of solar still: A review, *Heat Transfer* 2020;49(6):3758–3778. DOI: 10.1002/htj.21799.

[15] M. Benghanem, A. Mellit, M. Emad, A. Aljohani. Monitoring of solar still desalination system using the internet of things technique. *Energies* 2021;14(21):6892.

[16] K. Assalaou, L. Bouhouch, L. Elmahni. Development and optimization of a photovoltaic pumping irrigation system. *International Journal of Advanced Research* 5(4):513–521.

CHAPTER 7

Intelligent Temperature and Humidity Control System

Leveraging AI for Precise Climate Management

Najlaa Ben-Lhachemi

University Mohammed V in Rabat, Morocco

Mohammed Benchrifa

University Mohammed V in Rabat, Morocco
Ibn Tofaïl University—Kenitra-University Campus, Kenitra, Morocco

Souad Nasrdine, Jamal Mabrouki, and Miloudia Slaoui

University Mohammed V in Rabat, Morocco

7.1 INTRODUCTION

Industry 4.0 was founded on the principles of the first three industrial revolutions, which involved, respectively, steam power, electricity, and computers. Advanced technologies such as the Internet of Things (IoT), Artificial Intelligence (AI), cloud computing, and robotics are being integrated in this fourth industrial revolution to enable machine-to-machine communication and autonomous decision-making.

One such technology is AI, which learns and develops, anticipates, interacts with other systems, and ultimately makes judgments that are more accurate by evaluating a massive amount of data that humans are unable to do so and with a smaller margin of error than they do.

AI can help the agricultural sector increase productivity and efficiency (5). One of the most promising agricultural solutions is the use of automated smart greenhouses because humans lack the knowledge and expertise necessary to maintain them effectively and regulate plant growth conditions. Various AI technologies, including robotic systems, bio-inspired algorithms, energy management, machine path planning, drones, and image processing, are used in smart greenhouses, according to a study by Chrysanthos Maraveas.

DOI: 10.1201/9781003436218-7

Commercially feasible AI technologies for agriculture have proliferated, offering farmers larger returns on investment, greater crop yields, and improved resource efficiency [1].

By utilizing the information gathered by sensors and a web application for the best remote control, AI may be used to remotely manage plant growth, greenhouse climate, and nutrient levels. With the use of these capabilities, crop parameters may be tracked and changed in real time to increase output and decrease losses brought on by disease and changing weather.

In another study, Axel Escamilla-Garcia et al. investigated the use of artificial neural networks (ANN) in greenhouse technology and highlighted their potential for integration with other 4.0 technologies such as IoT, machine learning (ML), image analysis, and Big Data. At the moment, there is a lot of research on monitoring methods and approaches to manage the environment of a greenhouse. According to the report, additional ANN development study could enhance their utility in smart agriculture and greenhouse production [2].

At Wageningen University in 2018, five worldwide teams competed in a contest to apply AI algorithms to regulate the climate, irrigation, and crop growth in high-tech greenhouses. In their presentation of team findings, Silke Hemming et al. evaluate how different climates affect crop development and discuss lessons gained for improving crop yield and financial performance in the future [3].

In this work, we will use AI to our advantage, and to remotely control the temperature and humidity inside a greenhouse. We will propose the processes to follow for optimal control, while allowing the farmer to monitor and change the parameters remotely if necessary while connected to the internet.

7.2 LITERATURE REVIEW

7.2.1 AI Applications in Greenhouse Agriculture

Artificial Intelligence (AI) in greenhouse farming has the potential to revolutionize the agricultural sector by raising crop yields, lowering costs, and boosting productivity.

Bhanuka Senavirathne et al. divided the subsystems connected to greenhouses into major categories in their study. One of these categories was temperature regulation, which is essential in maintaining the greenhouse at the right temperature for plant growth. Controlling humidity is the next step since this helps to prevent fungal infections and manage plant transpiration. To give plants the nutrients they need to flourish, the identification of optimal nutrition is also crucial. Pumps are used to water plants and supply them with the nutrients they require. In addition to maintaining temperature and humidity, air ventilation is essential for regulating the concentration of carbon dioxide. Finally, artificial light is utilized to lengthen the period of time that plants are exposed to light and to supply sufficient light for plant growth [4].

Steps can be used to outline the essential components for effectively and efficiently controlling greenhouse subsystems. Data collection utilizing sensors and actuators should come first. The sensors collect data on a variety of variables, including temperature, humidity, air quality, and others. These settings are controlled by actuators to keep the environment ideal for plant development. The performance of the greenhouse is then predicted using the analysis of the acquired data. This makes it possible to make judgments regarding the greenhouse's environment that help optimize growing conditions for the highest crop production and quality possible. Finally, it is advised to use a web application to remotely monitor and control the greenhouse. With the help of this application, farmers may remotely monitor greenhouse conditions, make decisions in real time to optimize growing conditions, and control various greenhouse subsystems, including irrigation, lighting, cooling, and ventilation.

7.2.2 Limitations, Challenges, and Future Directions

There may be certain difficulties when implementing AI in agriculture, particularly in greenhouses. Self-learning greenhouse construction is costly, especially given the high cost of more precise sensors and controllers. Additionally, a lot of accurate data must be available in order to create this system. However, because so many technologies have not yet reached the commercial stage, the adoption of agricultural technologies is gradual. The majority of other technologies have not yet been put into practice in the field, with the exception of sensors, cameras, and some robots. The development of cameras and sensors to improve their accuracy and effectiveness is ongoing. Robots are typically faster and more accurate than people, yet they cannot replace them entirely. Younger farming generations are more receptive to technology advancement than older generations who have had less exposure to it.

Additionally, it is crucial that scientists and engineers collaborate with farmers and industry experts to design and implement solutions that address their needs and are both technically and financially viable.

To create AI models that can account for the complexity of agricultural ecosystems, including interactions between plants, insects, bacteria, and soil, research is required. Additionally, researchers can think about using and creating technologies inspired by nature for soil treatment, automated harvesting, and application of pesticides and fertilizers. For farmers, these tools are more precise, inexpensive, and simple to use.

Using AI-based technology in agriculture can assist the industry meet its sustainability, productivity, and profitability objectives.

7.3 MATERIALS AND METHODS

Remote climate control systems for greenhouses typically contain a number of common elements: sensors, actuators, a microprocessor, a user interface, and a wireless communication system. Actuators are used to control them by opening or closing fans or vents, while sensors are used to measure the various environmental conditions inside the greenhouse, such as temperature and humidity. The system's brain, the microcontroller, enables processing of sensor data and actuator control following set temperature settings. Remote monitoring and system control are made possible via the user interface, typically via a website or mobile app. The various system components may communicate with one other wirelessly and send each other user commands and temperature data as mentioned in Figures 7.1 and 7.2.

Sensors are tools that gauge many aspects of the environment, including temperature, humidity, light, air quality, etc. The ambient conditions in the greenhouse can be changed in real time using the information gathered by the sensors. On the basis of information gathered by sensors, actuators are devices that regulate the greenhouse's ambient conditions. The data gathered by the sensors and the commands issued by the actuators can be transferred between the control system and the greenhouse thanks to communication

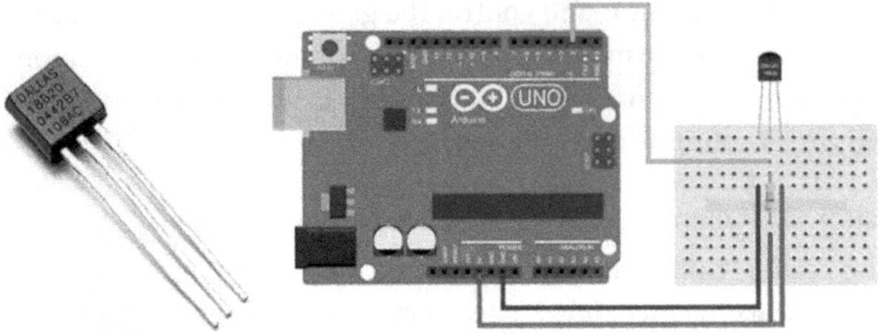

FIGURE 7.1 Temperature sensor with Arduino [5].

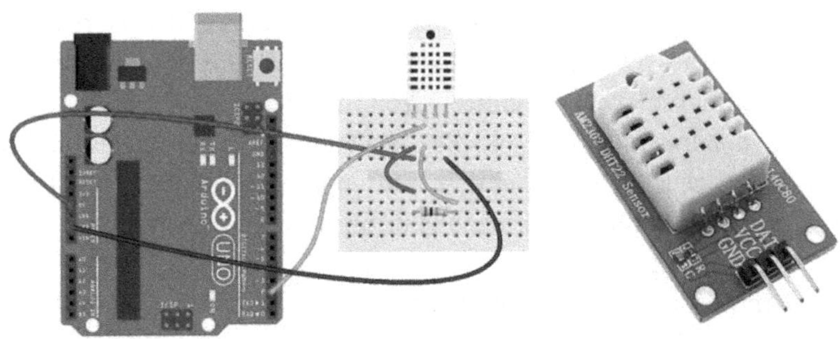

FIGURE 7.2 Humidity sensor with Arduino [6].

networks. The numerous gadgets can be linked together via wireless technologies like Wi-Fi or Bluetooth. Control systems are computer programs that enable monitoring and management of the various greenhouse equipment. Additionally, they can be used to gather data, evaluate it, and build control plans based on the requirements of certain plants.

7.4 RESULTS AND DISCUSSION

7.4.1 Proposal of an Intelligent System for the Control of a Greenhouse

Using the sensors, the system (Figure 7.3) keeps track of the temperature and humidity levels inside the greenhouse. Artificial intelligence (AI) is used to evaluate acquired data and find trends and patterns that may reveal issues or possibilities to improve greenhouse environmental conditions. The microprocessor sends a command to the actuators to regulate the temperature or humidity using the cooling or heating system in the case of temperature and the humidifier or dehumidifier in the case of humidity if they go above the predefined limits. The control system continuously assesses the greenhouse's environmental conditions and modifies the instructions given to the actuators in response to the information gathered.

It is possible to integrate a user interface to allow the user to monitor the temperature and humidity in real time and to modify the parameters remotely according to the needs. The user may monitor and manage the system from anywhere with an internet connection

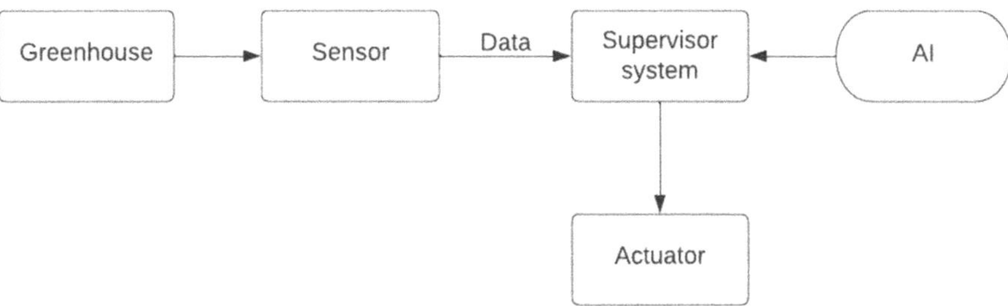

FIGURE 7.3 Schematization of the proposed system.

thanks to the wireless communication system, which makes remote monitoring convenient for farmers.

The difference between the greenhouse temperature and the intended temperature, which is the input to the greenhouse temperature control process shown in Figure 7.4, defines the kind of actuator (heating or cooling) selected by the supervisor controller. Similar to Figure 7.4, the difference between the actual and desired humidity is used to control the humidity in Figure 7.5. The dehumidifier and humidifier are acting as actuators in this instance.

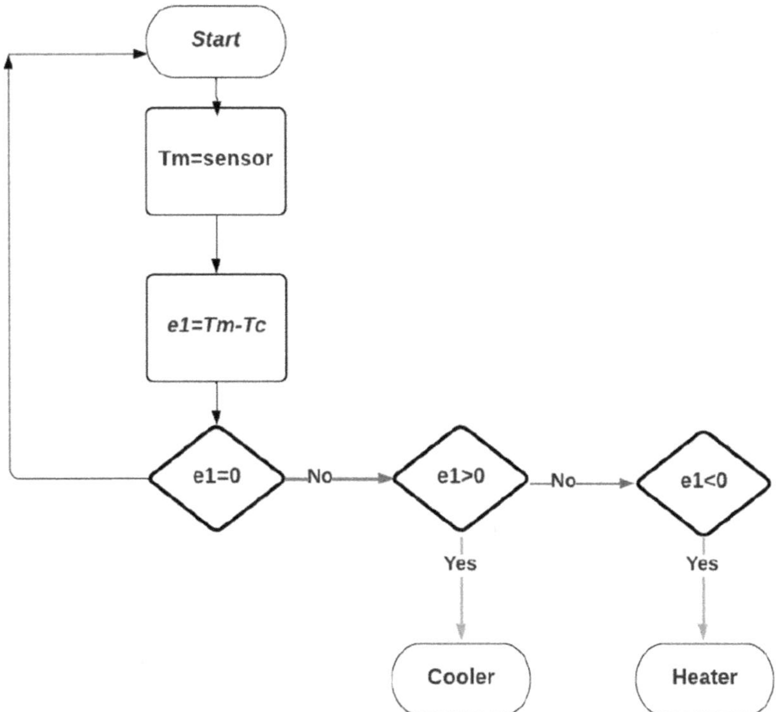

FIGURE 7.4 Flowchart showing the temperature control process.

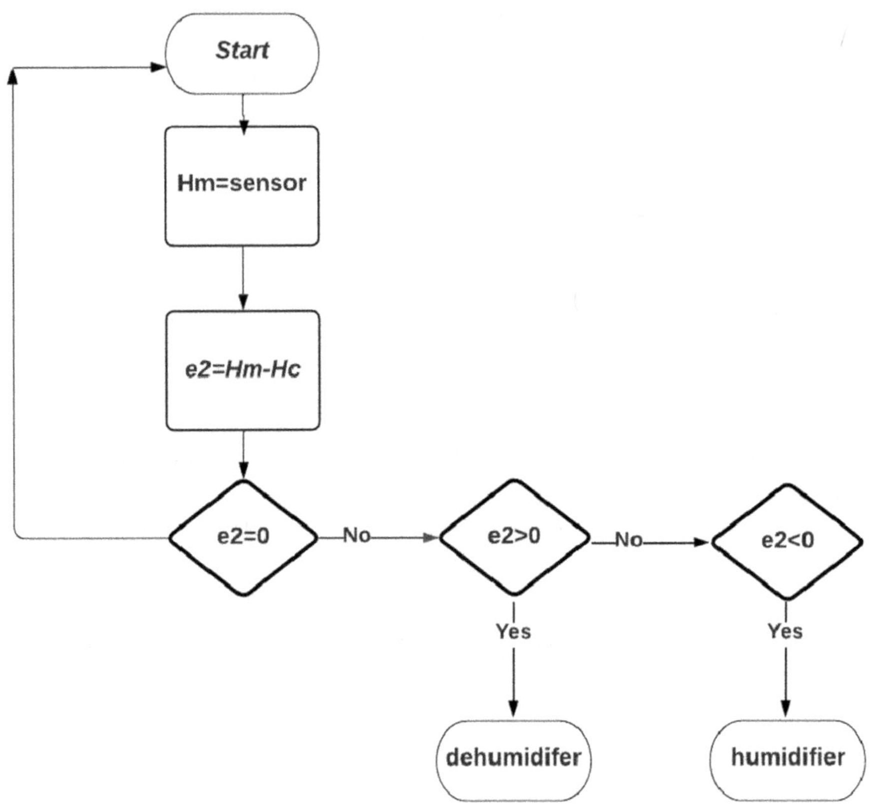

FIGURE 7.5 Flow chart showing the humidity control process.

7.5 CONCLUSION

Despite population expansion, using AI in agriculture may be a way to guarantee food security. Artificial Intelligence may assist farmers in optimizing their production and generating more food with fewer resources by offering precise weather forecasts, guidance on the management of soil and crops, and assistance in preventing plant diseases and pests. This could contribute to reducing environmental impact and guaranteeing the long-term sustainability of agricultural production while meeting the increasing global demand for food. Furthermore, Artificial Intelligence enables informed decision-making by providing accurate precise and current data about crops, soils, weather, and markets, which can assist farmers in maximizing their yields and preventing losses.

In this study, we present a system for managing the climate within a greenhouse using Artificial Intelligence. AI can alter the environment in real time to improve plant development by employing sensors to assess humidity and temperature levels. For instance, the AI can alter ventilation to boost the flow of fresh air if CO_2 levels are too low or can activate fans to circulate air if humidity levels are too high. Similarly, if the temperature rises too much, the AI can activate cooling devices to keep the temperature at a desirable level. By maximizing the utilization of resources, this can assist farmers in increasing yields, minimizing losses, and conserving energy. Additionally, growers can respond rapidly to possible issues like disease or pests by monitoring greenhouse conditions in real time, which can assist lower losses and improve crop quality. In the end, gardeners aiming to maximize productivity and raise the caliber of their products may find employing AI to regulate greenhouse climatic conditions to be a useful tool.

REFERENCES

[1] C. Maraveas, "Incorporating Artificial Intelligence Technology in Smart Greenhouses: Current State of the Art", *Appl. Sci.*, vol. 13, no. 1, January 2023, doi: 10.3390/app13010014.

[2] A. Escamilla-García, G. M. Soto-Zarazúa, M. Toledano-Ayala, E. Rivas-Araiza, and A. Gastélum-Barrios, "Applications of Artificial Neural Networks in Greenhouse Technology and Overview for Smart Agriculture Development", *Appl. Sci.*, vol. 10, no. 11, January 2020, doi: 10.3390/app10113835.

[3] S. Hemming, F. de Zwart, A. Elings, I. Righini, and A. Petropoulou, "Remote Control of Greenhouse Vegetable Production with Artificial Intelligence—Greenhouse Climate, Irrigation, and Crop Production", *Sensors*, vol. 19, no. 8, January 2019, doi: 10.3390/s19081807.

[4] B. Senavirathne and C. Gunerathne, "Greenhouse Automation with Artificial Intelligence and Industry 4.0 Integration", In *Proceedings of the 12th International Research Conference* (pp. 487–494), October 2019.

[5] F. Nawazi, "Interfacing DS18B20 1-Wire Digital Temperature Sensor with Arduino UNO", *Circuits DIY*, 21 March 2022. https://www.circuits-diy.com/interfacing-ds18b20-1-wire-digital-temperature-sensor-with-arduino-uno/ (accessed 24 April 2023).

[6] A. Khan, "Humidity Sensor with Arduino", *Circuits DIY*, 9 March 2022. https://www.circuits-diy.com/humidity-sensor-with-arduino/ (accessed 24 April 2023).

CHAPTER 8

Interaction between the Microalgae Cultivation System and Its Effects on Nutrient Removal from Wastewater

Khadija El-Moustaqim

Ibn Tofaïl University—Kenitra-University Campus, Kenitra, Morocco

Mohammed Benchrifa

University Mohammed V in Rabat, Morocco
Ibn Tofaïl University—Kenitra-University Campus, Kenitra, Morocco

Jamal Mabrouki

University Mohammed V in Rabat, Morocco

Driss Hmouni

Ibn Tofaïl University—Kenitra-University Campus, Kenitra, Morocco

8.1 INTRODUCTION

Water is a very limited natural resource in arid and semi-arid regions [1]. The existing resources are threatened by pollution from urban and industrial discharges to receptor environments [2–5]. These releases may contain many substances, both solid and dissolved, as well as many pathogenic microorganisms, which threaten the quality of the environment as a whole [6–8]. The water crisis is seen as one of the main global problems and threats, even if sufficient water and land resources are available [9]. According to the United Nations World Water Development Report (2014), more than two million tons of wastewater, agricultural and industrial waste are dumped untreated into lakes, rivers, and other bodies of water in developing countries, ending up polluting the usable water supply [10]. Wastewater from homes, industries, businesses, and institutions contains many substances that may pose

DOI: 10.1201/9781003436218-8

risks to human health and the environment [11]. Domestic wastewater, on the other hand, comes from lavatories as well as household water (kitchen, bathroom, laundry, and water generated by certain household appliances other than a lavatory) [12]. Among the effluents found in the wastewater are synthetic chemical compounds that are difficult to biodegrade, and mutagenic and/or carcinogenic toxicants from these effluents can be found at all levels of the food chain when they are discharged into receptor environments [13]. It is therefore necessary to find an alternative water treatment [14]. In recent years, countless biotechnological applications have emerged, such as bioremediation, which could allow wastewater treatment without all these disadvantages [15]. This process involves the use of microorganisms, usually bacteria or fungi, to remove pollutants from the soil and water [16]. The art of using algae (either macro- or microalgae) in the removal, biotransformation or mineralization of various nutrients, heavy metals and xenobiotics from wastewater and residual air carbon dioxide is known as phycoremediation [17–19]. During this treatment, carbon, nitrogen, phosphorus and other salts are used by algae as nutrients, wastewater or air as appropriate [20, 21]. Algae play key roles in the biological treatment of wastewater by lagoons, even as oxygen providers through the photosynthetic process [22]. Thus, they promote the oxidation of organic matter by associating in symbiotic form with bacteria [23]. They can even contribute directly to the elimination of certain organic derivatives and act as bio-absorbents contributing to the elimination of heavy metals and other toxic products [24].

The main objective of this work is to evaluate the effectiveness of the treatment of domestic wastewater through the use of microalgae as a technique of water purification studies, as well as the interest of their use for the renewable energy market and for the phycoremediation sector in order to achieve the purification of the nutrients and metals contained in the wastewater.

8.2 MATERIALS AND METHODS

8.2.1 Microalgae Culture System

The cultivation of microalgae using sunlight can be carried out either in open-air tanks or using tube-shaped photobioreactors, rectangles, continuous stirring reactors or other forms [18]. Microalgae are able to grow quickly and can be grown by different methods and under different conditions. This biomass is shown to be an attractive raw material for the rapid generation of biomolecules, which can supplement the nutritional, pharmaceutical, cosmetic and energy needs of the population [25, 26]. These photosynthetic microorganisms develop very quickly, by transforming about 10% of solar energy into biomass with a duplication of between 0.3 and 3 times more and an estimated theoretical yield of about 77 g/biomass/m^2/day [27, 28].

a. **Open system**:
   Open cultures such as ponds and uncovered tanks (indoors or outdoors) are more easily contaminated than closed culture vessels such as tubes, flasks, carboys, bags, etc. Axenic culture can also be chosen, using algal cultures free of any foreign organisms such as bacteria; this is expensive and difficult, however, as it requires strict sterilization of all glassware, culture media and containers to avoid contamination [29]. The use of open systems as a microalgae culture method is quite common.

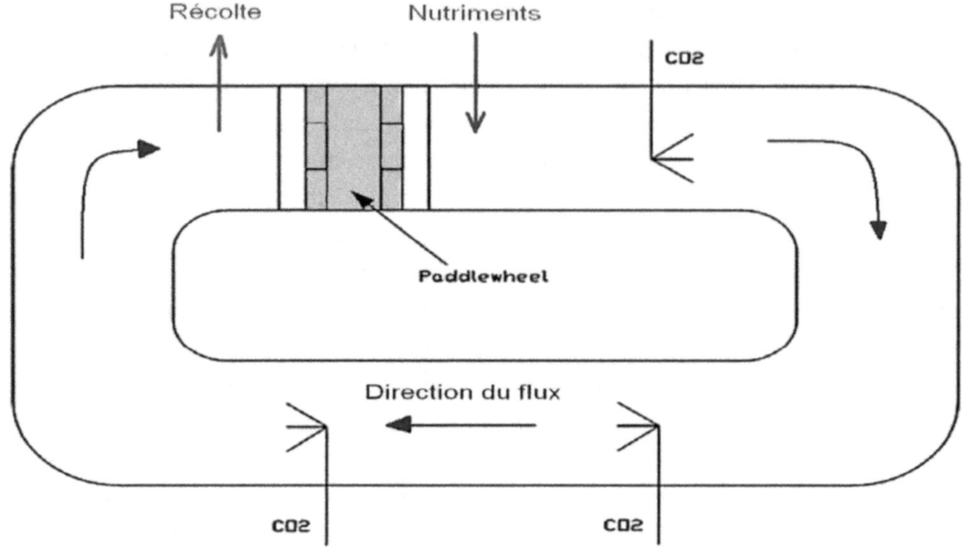

FIGURE 8.1 Open system diagram.

Open systems come in many different forms, each with certain advantages and disadvantages. The types of systems that are currently used in research and industry include pipe ponds, large shallow ponds, circular ponds and closed systems [30]. Figure 8.1 shows a representative diagram of an open culture system.

b. **Closed system**:
Photobioreactors are reactors made from transparent materials. Their design is based on the illuminated surface, the efficiency of the mixture and the control of the culture parameters (temperature, carbon dioxide and oxygen content, pH), to achieve maximum productivity. Closed systems have been designed to address pool problems [31]. They offer a closed culture environment; they protect the culture from direct contamination; and they allow a better control of the conditions of cultivation. The temperature is controlled effectively, the access to light is increased compared to the basin, the evaporation of the growing medium is minimized, the supply of CO_2 is facilitated, and its losses are limited [32]. The photobioreactor allows the efficient arrangement of light but also the removal of oxygen produced by photosynthesis [33]. For better control of the process, a possible solution is to carry out closed crops. The principle of closed systems used for microalgal culture is to put microalgae in contact with natural light, combining both through a transparent material [29].

8.2.2 Advantages and Disadvantages

Microalgae culture systems have both advantages and disadvantages depending on the desired products where they are classified in Table 8.1.

Despite this, the success of each small- or large-scale cropping system depends on many parameters, such as the efficient intensity of the light source, the optimal transfer of gas and liquid, the ease of use, low level of contamination, low production, minimum area, nutrient maximization, and temperature control (Figure 8.2).

TABLE 8.1 Advantages and Disadvantages of Microalgae Culture Systems

Advantages	Ref.	Disadvantages	Ref.
Open system			
• A low cost • Maintenance • Construction • Configuration is easy to complete	[34–37]	• Difficult to control growing conditions • Threats of contamination • Evaporative losses	[38–40]
Closed system			
• Control growing conditions and growth parameters (pH, CO_2, O_2) • Avoid evaporation to reduce CO_2 losses • Achieve higher microalgae densities	[41–44]	• Overheating • Biofouling • The accumulation of oxygen • The high cost of construction	[39, 45, 46]

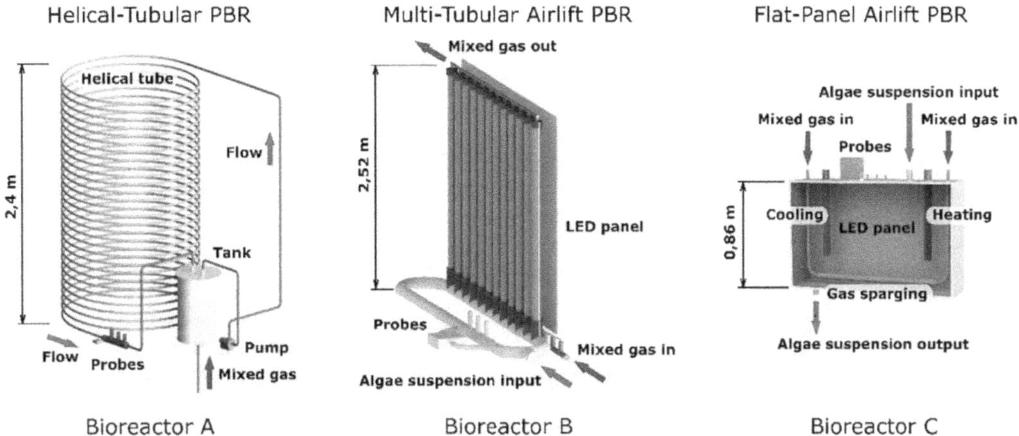

FIGURE 8.2 Refers to photobioreactor perspective.

8.2.3 Interaction of Microalgae with Nutrients

It is important to note that photosynthetic organisms have evolved in a constantly changing world. Photosynthetic organisms collect energy from sunlight through pigment-protein complexes located in membranes in order to promote energy transfer [47]. Figure 8.3 shows the types and broad classes of microalgae, with characteristics:

Nutrients such as carbon, nitrogen, and phosphorus play an important role in cell metabolism and the biochemical composition of microalgae [48]. In addition, for microalgae growth, there must be the presence of trace elements, especially metals, such as Mg, Ca, Mn, Zn, Cu, and Mo [49, 50]. Microalgae play an important role in nitrogen fixation and uptake. They include cyanobacteria that are capable of binding atmospheric molecular nitrogen (N_2-N) and converting it to ammonia nitrogen (NH_3-N), which can be excreted into the environment by the following equation [51]:

$$N_2 + 8H^+ + 8_{e^-} + 16 + ATP \xrightarrow{\text{Nitrogenase}} 2NH_3 + H_2 + 16\,ADP + 16P_i \tag{8.1}$$

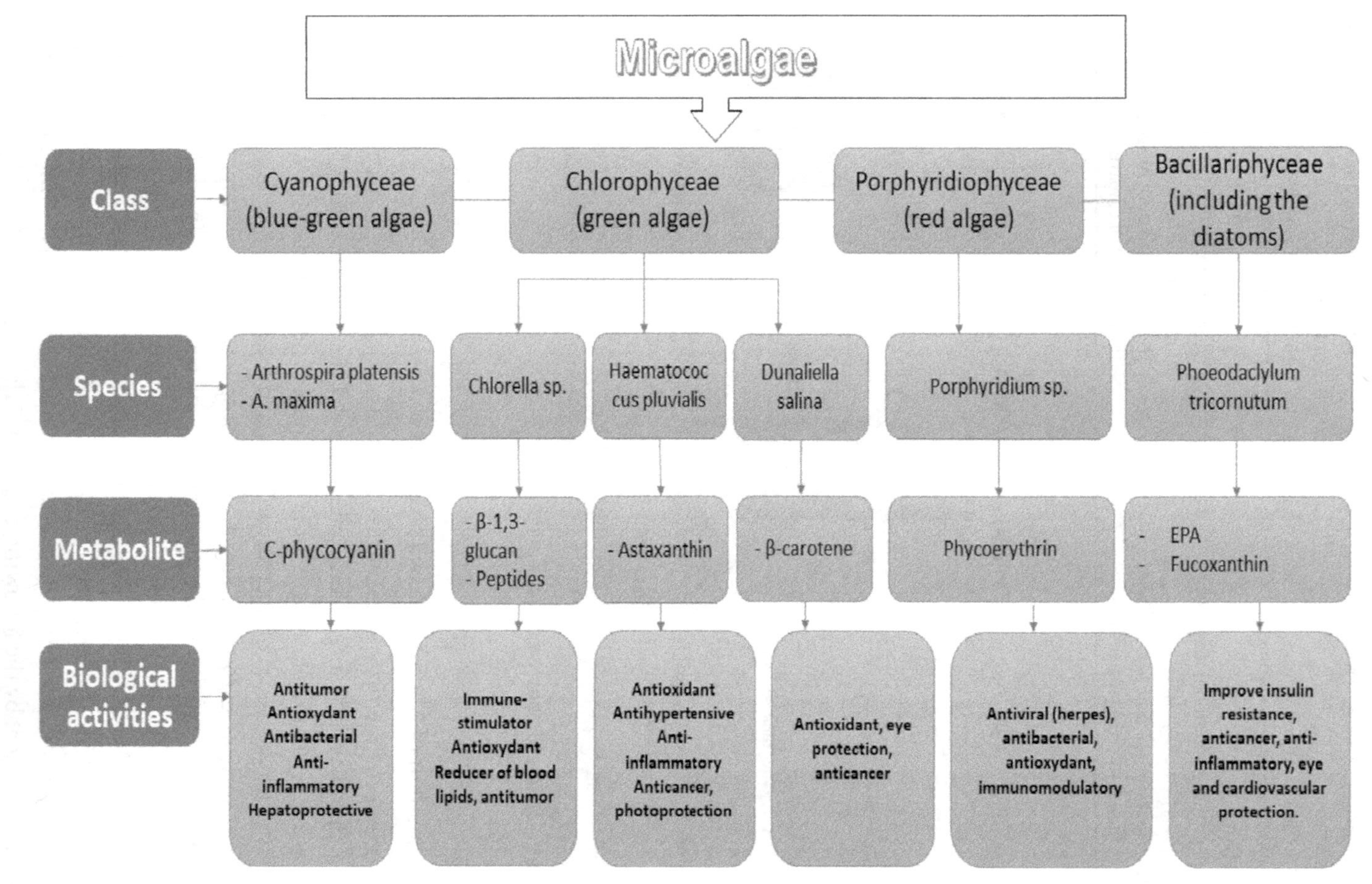

FIGURE 8.3 Classification of the main types of microalgae, their species and their biological activities.

Eukaryotic microalgae, in turn, are able to assimilate fixed nitrogen such as NH_4-N, NO_3-N and nitrate reductase and nitrite reductase according to the following two equations [52]:

$$NO_3^- + 2H^+ + 2_e^- \xrightarrow{\text{Nitrate reductase}} NO_2^- + H_2O \quad (8.2)$$

$$NO_2^- + 8H^+ + 6_e^- \xrightarrow{\text{Nitrite reductase}} NH_4^+ + 2H_2O \quad (8.3)$$

As long as NH_4-N uptake does not require reduction steps, it is thought to be the preferred form of nitrogen for microalgae. Therefore, in addition to the absorption of microalgae, the removal of NH_4-N can occur in response to an increase in pH and temperature, where large amounts of NH_4-N can be volatilized [53, 54].

8.3 RESULTS AND DISCUSSIONS

8.3.1 Wastewater Treatment Techniques Using Microalgae

For an example of the use of microalgae as a wastewater treatment technique which contains a high level of nutrients from physicochemical analyses, we take the example of the work elaborated by Gharmouli et al., who studied the removal of organic pollutants contained in wastewater by *Arthrospira platensis.* Based on this study, they found a decrease in electrical conductivity and the absence of ammonia and nitrite in the medium with a decrease in nitrate value, as well as total nitrogen under a temperature of 32°C [55]. A comparative study on the use of microalgae-bacterial consortia in wastewater treatment to examine and compare microbial communities associated with a green algae consortium was conducted. The results show that a large part of the nitrogen and carbon has been removed, which opens up a potentially economically viable wastewater treatment process [56]. A very good result is obtained in the work of De La Noue et al., who carried out the study on an urban effluent by a culture of the species Scenedesmus on a modified medium, in which they noticed a depletion greater than 90% is obtained for 48 hours for nitrogen and phosphorus. It appears that the raw effluent provides a growth greater than the synthetic medium and comparable to the filtered effluent, which indicates the potential for Scenedesmus may be used as a tertiary treatment agent [57]. Also, the work of Fredereque proposes a new approach to the production of microalgae in industrial parks to treat a mixture of different wastewater that has allowed to climb the main genera that take over the algae when growing conditions change. Based on this study, he was able to demonstrate that the algae-bacteria consortium used is effective in removing certain nutrients from wastewater [58]. The application of wastewater contaminated with metals is studied by Samadani. In his study, he used green algae *Chlamydomonas reinhardtii* and *Chlamydomonas acidophila* to select the species most resistant to Cd. He found that, species *C. reinhardtii* showed a higher intracellular Cd accumulation capacity compared to *C. acidophila* [59].

8.3.2 Nutritional Value in the Species *Chlorella*

Biotechnology interest in microalgae and cyanobacteria has focused on their content in valuable metabolites that can be used to develop a wide range of different applications in fields such as medicine, industry, energy, agriculture, food, and others [60]. Most cosmetic actions, supported by the literature and claimed by the market, are, on the one hand, anti-aging, including moisturizing, antioxidant and lightening properties, and, on the other hand, anti-acne and healing properties [61].

8.4 CONCLUSION

Photosynthetic microorganisms are becoming established as one of the most important alternatives for implementing it in the techniques of wastewater treatment and noted a growing interest, in particular with regard to the microalgae that are gradually imposed in the scientific and industrial world. The interest in microalgae is currently growing as they are regarded as organisms capable of removing nutrients from wastewater. This work shows the importance of the use of microalgae in wastewater treatment. Therefore, in order to select the most appropriate species for the purpose of removing nutrients from the environment, the species' ability to accumulate and its tolerance against toxic effects must be studied.

REFERENCES

[1] D. Mani and C. Kumar, "Biotechnological advances in bioremediation of heavy metals contaminated ecosystems: an overview with special reference to phytoremediation", *Int. J. Environ. Sci. Technol.*, vol. 11, no. 3, pp. 843–872, April 2014, doi: 10.1007/s13762-013-0299-8.

[2] Y. A. Idrissi, A. Alemad, S. Aboubaker, H. Daifi, K. Elkharrim, and D. Belghyti, "Caractérisation physico-chimique des eaux usées de la ville d'Azilal, Maroc" (Physicochemical characterization of wastewater from Azilal city, Morocco), vol. 11, no. 3, p. 12, 2015.

[3] M. Gafsi, A. Kettab, S. Benmamar, and S. Benziada, Cas d'une pollution: l'eutrophisation dans les cours d'eau, p. 9, 2008.

[4] S. Zgheib, R. Moilleron, and G. Chebbo, « Flux et sources des polluants prioritaires dans les eaux urbaines en lien avec l'usage du territoire », *Tech. Sci. Méthodes*, vol. 4, pp. 50–62, 2011, doi: 10.1051/tsm/201104050.

[5] A. Rassam, A. Chaouch, B. Bourkhiss, M. Ouhssine, T. Lakhlifi, and M. Bourkhiss, « Caractéristiques physico-chimiques des eaux usées brutes de la ville d'Oujda (Maroc) », *Les technologies de laboratoire*, vol. 7, no. 28, 2012.

[6] M. Makhoukh, M. Sbaa, and A. Berrahou, « Contribution à l'etude physico-chimique des eaux superficielles de l'oued moulouya (Maroc oriental) », 2011.

[7] F. Dimane, K. Haboubi, I. Hanafi, and A. El Himri, « Étude de la Performance du Dispositif de Traitement des Eaux Usées par Boues Activées de la ville d'Al- Hoceima, Maroc », *Eur. Sci. J. ESJ*, vol. 12, no. 17, p. 272, June 2016, doi: 10.19044/esj.2016.v12n17p272.

[8] H. Elaouani, H. Haffad, K. Jaafari, N. Elbada, S. Mailainine, and K. Benkhouja, « Evaluation de la qualité physico-chimique des rejets liquides industriels de la zone industrielle d'elmarsa laayoune ». *Revue de l'entrepreneuriat et de l'innovation*, vol. 2, no. 7.

[9] X. Wan, M. Lei, and T. Chen, « Cost–benefit calculation of phytoremediation technology for heavy-metal-contaminated soil », *Sci. Total Environ.*, vol. 563–564, pp. 796–802, Sept. 2016, doi: 10.1016/j.scitotenv.2015.12.080.

[10] D. W. Schindler et al., « Eutrophication of lakes cannot be controlled by reducing nitrogen input: results of a 37-year whole-ecosystem experiment », *Proc. Natl. Acad. Sci.*, vol. 105, no. 32, pp. 11254–11258, 2008.

[11] M. L. O. Diakité, « Traitement des eaux usées municipales par phanerochaete chrysosporium », École de technologie supérieure, 2016.

[12] « 25 ans d'assainissement des eaux usées industrielles au Québec : un bilan ». https://www.environnement.gouv.qc.ca/eau/eaux-usees/industrielles/chapitre1_b.htm (accessed 24 May 2023).

[13] P. J. Hodgson, *Étude de la dégradation des acides résiniques par le champignon Phanerochaete chrysosporium*. National Library of Canada= Bibliothèque nationale du Canada, Ottawa, 2000.

[14] S. N. Singh and R. D. Tripathi, *Environmental bioremediation technologies*. Springer Science & Business Media, 2007.

[15] E. J. Olguín and G. Sánchez-Galván, « Heavy metal removal in phytofiltration and phycoremediation: the need to differentiate between bioadsorption and bioaccumulation », *New Biotechnol.*, vol. 30, no. 1, pp. 3–8, 2012.

[16] W. Mulbry, J. Reeves, Y. Liu, Z. Ruan, and W. Liao, « Near-and mid-infrared spectroscopic determination of algal composition », *J. Appl. Phycol.*, vol. 24, pp. 1261–1267, 2012.

[17] D. E. Gharmouli, A. Abdaoui, and A. Souide, « La Bioremedition des eaux usées par des microalgues dans la région d'El Oued », 2022.

[18] M. Sadi, « Les micro algues: un défi prometteur pour des biocarburants propres », *Rev. Energ. Renouvelables SIENR*, vol. 12, pp. 195–202, 2012.

[19] B. Sialve and J.-P. Steyer, « Les microalgues, promesses et défis », *Innov. Agron.*, vol. 26, pp. 25–39, 2013.

[20] J. Person, « Algues, filières du futur », *Livre Turquoise Adebiotech*, 2010.

[21] M.-È. Boileau, « Évaluation du potentiel d'utilisation d'une eau usée industrielle comme substrat de culture pour des microalgues d'eau douce dans une optique de production de biocarburants de 3e génération », Université de Sherbrooke, 2015.

[22] C. Sili, G. Torzillo, and A. Vonshak, « Arthrospira (Spirulina) », *Ecol. Cyanobacteria II Their Divers. Space Time*, pp. 677–705, 2012.

[23] S. A. Razzak, S. A. M. Ali, M. M. Hossain, and H. deLasa, « Biological CO_2 fixation with production of microalgae in wastewater – A review », *Renew. Sustain. Energy Rev.*, vol. 76, pp. 379–390, 2017.

[24] S. Wang, J. Liu, C. Li, and B. M. Chung, « Efficiency of Nannochloropsis oculata and Bacillus polymyxa symbiotic composite at ammonium and phosphate removal from synthetic wastewater », *Environ. Technol.*, vol. 40, no. 19, pp. 2494–2503, 2019.

[25] K. Samhat, « *Contribution à l'optimisation de la production d'astaxanthine en photobioréacteur à partir de la microalgue Haematococcus pluvialis* », Nantes Université; Université Libanaise, 2023.

[26] N. Daghouche and B. S. Grama, « Valorisation des sous-produits d'origines algales », 2021.

[27] M. I. Khan, J. H. Shin, and J. D. Kim, « The promising future of microalgae: current status, challenges, and optimization of a sustainable and renewable industry for biofuels, feed, and other products », *Microb. Cell Factories*, vol. 17, no. 1, pp. 1–21, 2018.

[28] J.-P. Cadoret and O. Bernard, « La production de biocarburant lipidique avec des microalgues: promesses et défis », *J. Société Biol.*, vol. 202, no. 3, pp. 201–211, 2008.

[29] L. Barsanti and P. Gualtieri, *Algae: anatomy, biochemistry, and biotechnology*. CRC Press, 2022.

[30] F. Doré-Deschênes, *Utilisation des microalgues comme source d'énergie durable*. Centre universitaire de formation en environnement, Université de Sherbrooke, 2009.

[31] L. Brennan and P. Owende, « Biofuels from microalgae—A review of technologies for production, processing, and extractions of biofuels and co-products », *Renew. Sustain. Energy Rev.*, vol. 14, no. 2, pp. 557–577, 2010.

[32] A. Contreras, F. García, E. Molina, and J. C. Merchuk, « Interaction between CO_2-mass transfer, light availability, and hydrodynamic stress in the growth of Phaeodactylum tricornutum in a concentric tube airlift photobioreactor », *Biotechnol. Bioeng.*, vol. 60, no.3, pp. 317–325, 1998.

[33] R. W. Babcock, J. Malda, and J. C. Radway, « Hydrodynamics and mass transfer in a tubular airlift photobioreactor », *J. Appl. Phycol.*, vol. 14, pp. 169–184, 2002.

[34] L. Novoveská, A. K. Zapata, J. B. Zabolotney, M. C. Atwood, and E. R. Sundstrom, « Optimizing microalgae cultivation and wastewater treatment in large-scale offshore photobioreactors », *Algal Res.*, vol. 18, pp. 86–94, 2016.

[35] J. Wolf et al., « Multifactorial comparison of photobioreactor geometries in parallel microalgae cultivations », *Algal Res.*, vol. 15, pp. 187–201, 2016.

[36] T. M. Mata, A. A. Martins, and N. S. Caetano, « Microalgae for biodiesel production and other applications: a review », *Renew. Sustain. Energy Rev.*, vol. 14, no. 1, pp. 217–232, 2010.

[37] L. Moreno-Garcia, K. Adjallé, S. Barnabé, and G. S. V. Raghavan, « Microalgae biomass production for a biorefinery system: recent advances and the way towards sustainability », *Renew. Sustain. Energy Rev.*, vol. 76, pp. 493–506, 2017.

[38] M. A. Borowitzka, *Open pond culture systems. Algae for biofuels and energy. MA Borowitzka and NR Moheimani.* Springer, 2013.

[39] A. P. Carvalho, L. A. Meireles, and F. X. Malcata, « Microalgal reactors: a review of enclosed system designs and performances », *Biotechnol. Prog.*, vol. 22, no. 6, pp. 1490–1506, 2006.

[40] Y. Chisti, « Biodiesel from microalgae beats bioethanol », *Trends Biotechnol.*, vol. 26, no. 3, pp. 126–131, 2008.

[41] O. Pulz and K. Scheibenbogen, « Photobioreactors: design and performance with respect to light energy input », *Bioprocess Algae React. Technol. Apoptosis*, vol. 59, pp. 123–152, 2006.

[42] C. U. Ugwu, H. Aoyagi, and H. Uchiyama, « Photobioreactors for mass cultivation of algae », *Bioresour. Technol.*, vol. 99, no. 10, p. 4021–4028, 2008.

[43] J. U. Grobbelaar, « Factors governing algal growth in photobioreactors: the "open" versus "closed" debate », *J. Appl. Phycol.*, vol. 21, pp. 489–492, 2009.

[44] R. Harun, M. Singh, G. M. Forde, and M. K. Danquah, « Bioprocess engineering of microalgae to produce a variety of consumer products », *Renew. Sustain. Energy Rev.*, vol. 14, no. 3, pp. 1037–1047, 2010.

[45] O. Pulz, « Photobioreactors: production systems for phototrophic microorganisms », *Appl. Microbiol. Biotechnol.*, vol. 57, pp. 287–293, 2001.

[46] S. Amin, « Review on biofuel oil and gas production processes from microalgae », *Energy Convers. Manag.*, vol. 50, no. 7, pp. 1834–1840, 2009.

[47] A. Norici, C. Gerotto, and J. A. Raven, « Variabilité environnementale et son contrôle de la productivité », *Planète Bleue Photosynth. Rouge Verte Product. Cycle Carbone Dans Écosystèmes Mar.*, p. 229, 2023.

[48] J.-L. Fan, Y.-J. Zhang, and B. Wang, « The impact of urbanization on residential energy consumption in China: An aggregated and disaggregated analysis », *Renew. Sustain. Energy Rev.*, vol. 75, pp. 220–233, 2017.

[49] J. U. Grobbelaar, « Algal nutrition: mineral nutrition. », *Handb. Microalgal Cult. Biotechnol. Appl. Phycol.*, pp. 97–115, 2004.

[50] A. Kumar et al., « Enhanced CO2 fixation and biofuel production via microalgae: recent developments and future directions », *Trends Biotechnol.*, vol. 28, no. 7, pp. 371–380, 2010.

[51] T. Cai, S. Y. Park, and Y. Li, « Nutrient recovery from wastewater streams by microalgae: status and prospects », *Renew. Sustain. Energy Rev.*, vol. 19, pp. 360–369, 2013.

[52] C. L. Crofcheck, M. Monstross, E. Xinyi, A. P. Shea, M. Crocker, and R. Andrews, « Influence of media composition on the growth rate of Chlorella vulgaris and Scenedesmus acutus utilized for CO_2 mitigation », in *2012 Dallas, Texas, July 29–August 1, 2012*, American Society of Agricultural and Biological Engineers, 2012, p. 1.

[53] M. Wang, W. C. Kuo-Dahab, S. Dolan, and C. Park, « Kinetics of nutrient removal and expression of extracellular polymeric substances of the microalgae, *Chlorella* sp. and *Micractinium* sp., in wastewater treatment », *Bioresour. Technol.*, vol. 154, pp. 131–137, February 2014, doi: 10.1016/j.biortech.2013.12.047.

[54] Y. Su, A. Mennerich, and B. Urban, « Synergistic cooperation between wastewater-born algae and activated sludge for wastewater treatment: influence of algae and sludge inoculation ratios », *Bioresour. Technol.*, vol. 105, pp. 67–73, 2012.

[55] D. E. Gharmouli, A. Abdaoui, and A. Souide, « La Bioremedition des eaux usées par des micro-algues dans la région d'El Oued. », 2022 (accessed 27 May 2023). [Online.] Available at: http://dspace.univ-eloued.dz:80/xmlui/handle/123456789/12940

[56] M. Cinq-Mars, « *Caractérisation microbienne de deux consortiums de microalgues-bactéries cultivés en eaux usées industrielles et recherche de molécules d'intérêt* », Université du Québec à Trois-Rivières, 2021.

[57] J. de la Noüe, R. van Coillie, L. Brunel, and Y. Pouliot, « Traitement des eaux usées par culture de micro-algues: influence de la composition du milieu sur la croissance de Scenedesmus sp », *Ann Limnol-Int J Lim*, 1989, vol. 25, no. 3, pp. 197–203.

[58] F. Bélanger-Lépine, « Culture d'un consortium d'algues-bactéries dans des eaux usées industrielles pour l'obtention de produits biosourcés utilisables par les entreprises locales », Université du Québec à Montréal Université du Québec à Trois-Rivières, 2019.

[59] M. Samadani, « Étude comparative des effets de l'accumulation intracellulaire du cadmium chez les algues Chlamydomonas reinhardtii et Chlamydomonas acidophila », Université du Québec à Montréal, 2014.

[60] M. Martínez-Ruiz et al., « Microalgae bioactive compounds to topical applications products—A Review », *Molecules*, vol. 27, no. 11, January 2022, doi: 10.3390/molecules27113512.

[61] A. Benmhidi and B. S. Grama, « Valorisation et utilisation des microalgues en cosmétique », 2022 (accessed 27 May 2023). [Online.] Available at: http://localhost:8080/xmlui/handle/123456789/14341

CHAPTER 9

Study of Stability and Evolution of Milk Composition, Fat Globule Size and Acids in Raw Cow's Milk under Natural Conditions

Qisse Najat

University Mohammed V in Rabat, Morocco

Mohammed Benchrifa

University Mohammed V in Rabat, Morocco
Ibn Tofaïl University—Kenitra-University Campus, Kenitra, Morocco

Jamal Mabrouki

University Mohammed V in Rabat, Morocco

9.1 INTRODUCTION

In Morocco, milk is an essential part of the diet for many. It is consumed in a variety of forms, including liquid milk, dairy products, yoghurt and cheese. Milk is the integral product of the complete and uninterrupted milking of a healthy, well-nourished and not overworked dairy cow. It must be collected cleanly and contain no colostrum [1]. Cow's milk is a product with a number of nutritional qualities that explain why it is such an important part of the food pyramid. First and foremost, it is very high in calcium, a mineral essential for growth and bone health, as well as for a number of biological functions. It is one of the foods that provides the most calcium per serving.

Milk is a complex product; in-depth knowledge of its composition, the structural organization of its compounds and its physicochemical properties is essential to understanding milk processing, and the products obtained during the various treatments applied on an industrial

DOI: 10.1201/9781003436218-9

scale [2]. Milk is the basic raw material in the preparation of a wide range of fermented dairy products, representing a wide variety in terms of preparation, presentation and organoleptic quality [3]. Milk is a complex reaction medium whose primary role is to satisfy all the infant's nutritional needs. It mainly contains carbohydrates, fat, proteins and mineral salts. Other constituents found in milk in trace form include vitamins, enzymes and dissolved gases. The natural pH of milk is between 6.6 and 6.8, and is greatly influenced by its composition.

Reconstituted milk is obtained by mixing treated water with powdered milk. Reconstituted milk is said to be skimmed when powdered milk containing 1.25% fat per 100 g of powdered milk is used. Reconstituted milk is said to be whole, when powdered milk containing at least 26% fat per 100 g of powdered milk is used [4]. The major constituents of cow's milk are proteins, which are of particular interest as they are the main contributors to the technological properties of milk. Proteins account for 95% of the nitrogenous matter in milk, and can be divided into two groups: caseins and whey proteins. Non-protein nitrogen is made up of various substances such as urea, ammonia, uric acid, free amino acids and peptides. The study's main aim is to determine the microbiological and physicochemical quality of raw milk. As far as microbiological quality is concerned, the total aerobic mesophilic flora must be determined and counted. With regard to physicochemical quality, the following elements must be determined or measured: titratable acidity; storage temperature; fat content; alcohol test; protein content: total dry extract; pH.

9.2 LITERATURE REVIEW

As man's first food, milk is the only food that can claim universal status, at all times and in all places, at least for the first part of human life [1]. Milk is an organic food of considerable nutritional value, and its organized production dates back over ten thousand years. Since the 19th century, production has been steadily increasing, thanks to advances in veterinary medicine, the selection of high-performance breeds and breeding practices [2].

Milk is an opaque, matt-white, slightly yellowish alimentary liquid with a mild odor and sweet taste, secreted, after parturition, by the mammary glands of female mammals such as cows, goats and sheep, and intended for the feeding of newborn young animals. [3] All the names given to powdered milk, i.e. dehydrated or dry milk, refer to milk from which almost all the water has been removed [5]. This removal takes place after pasteurization and concentration at moderate temperature (71–75°C for 15–40 seconds), so the milk is finely pulverized in a drying tower heated by a stream of hot, dry air (approx. 150°C). The water evaporates, and the powder thus obtained undergoes the granulation process [4].

9.2.1 Chemical Composition of Milk

Milk is an important source of essential nutrients and a significant source of dietary energy, high-quality protein and fat. It can make a significant contribution to the recommended nutritional requirements for calcium, magnesium, selenium, riboflavin, vitamin B12 and pantothenic acid. Dairy products are nutritious foods that can be used to diversify plant-based diets. In populations with low fat intakes and limited access to other foods of animal origin, animal milk can play an important role in children's diets [6].

9.2.2 Nutritional Value of Milk and Regulations

Raw cow's milk contains all the nutrients required for the growth of young mammals: a liter of bovine milk contains around 50 g lactose, 32 g protein and 40 g fat. The energy potential of a liter of whole, semi-skimmed or skimmed milk is 2720 KJ, 2090 KJ and 1460 KJ respectively. Milk is an excellent source of calcium and phosphorus, as well as vitamins such as riboflavin, thiamine, cobalamin and vitamin A. It contains little iron and copper, little ascorbic acid and niacin, and relatively little vitamin D. It contains proteins rich in essential amino acid residues and minerals of nutritional interest (calcium and phosphorus), notably in the form of phosphates, citrates and chlorides of calcium, magnesium, potassium and sodium [7]

The standard published by AFNOR, entitled "Determination of the acidity titratable in chemistry", was published in Paris in 1995. It is part of the AFNOR series of standards concerning measurement methods in chemistry. This document describes the specific method for determining titratable acidity in chemical samples. Titratable acidity is a measure of the acidity of a solution, expressed in millimoles of acid per liter (mol/L) or as a percentage of acid. The method described in this chapter is based on the use of a titrant, usually a basic solution, which is added to the sample to react with the acid species present. Titratable acidity is determined by the quantity of titrant required to reach an equivalence point, where all acid species have reacted. This AFNOR standard is of significant importance for the harmonization of titratable acidity measurement methods in chemistry. It provides precise, recognized guidelines for laboratories and professionals in the chemical industry. Following this standard ensures the reliability and comparability of titratable acidity analysis results, facilitating the exchange of information between the various players in the chemical industry [8, 9].

9.3 MATERIAL AND METHODS

9.3.1 Sampling, Extraction and Storage

This study involved 4 samples of raw milk from 4 representative dairies in the P1, P2, P3 and P4 districts, at 3 different stages of reception. The samples were taken in the morning and placed in sterilized vials, which were then placed directly in freezers to prevent the temperature from rising.

It is not possible to pasteurize milk immediately upon receipt. Before it can be treated, milk has to be stored in storage tanks for several hours or days. Even with intensive refrigeration, it is difficult to avoid serious deterioration in its quality. So thermization is used to preheat milk to a lower temperature than pasteurization, to temporarily inhibit the growth of bacteria, especially pathogens. The milk is heated to 75°C, a combination of temperature and time that does not activate the phosphatase enzyme (used to check pasteurization efficiency and prevent the multiplication of aerobic spore-forming bacteria). After thermization, the milk must be rapidly cooled to 4°C or less, and must not be mixed with untreated milk. Many experts believe that thermization has a positive effect on certain spore-forming bacteria, as the heat treatment returns many spores to a vegetative state, enabling them to be destroyed during subsequent pasteurization of the milk. The milk, which is initially at a temperature of 4 to 8°C, is heated to 75°C in order to destroy a good proportion of micro-organisms and maintain the product at an acceptable bacteriological level until it is used. To achieve this, the raw milk is passed through a plate thermizer. The thermizer consists of three sections:

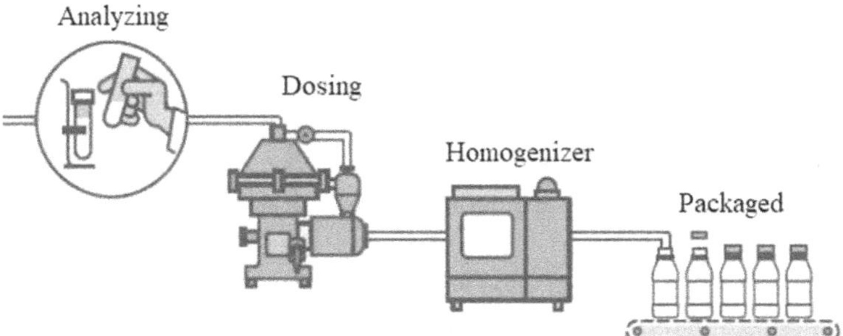

FIGURE 9.1 Receiving, preparing and unloading milk.

- Preheating: Raw milk, coming from the storage tank at a temperature of between 4 and 6°C, passes through a preheating section where it recovers heat from the hot milk thanks to a cold milk/hot milk circuit. The milk thus emerges at a temperature of 45°C, and is then sent to the skimmer before returning to the thermisor at the same temperature (Figure 9.1).
- Heating: In this section, the temperature of the milk is raised to 75°C using hot water via a counter-current plate heat exchanger. The milk then passes through a chamberer, where it is maintained at a temperature of 75°C for the required time.
- Cooling: This section cools the milk, which has already been precooled by indirect contact with the raw milk entering the thermisor. The milk is then subjected to final cooling to 6°C using chilled water.

The various stages of the thermisor allow the raw milk to be preheated, to reach a temperature of 75°C for partial destruction of the micro-organisms, and then to be rapidly cooled to maintain its quality before further use [10, 11].

a. **Physicochemical analysis**

In assessing milk stability, various parameters are taken into account for physicochemical analysis. The pH of raw milk is a crucial parameter, as significant fluctuations can indicate changes in its chemical composition, which can thereby affect its stability. The alcohol test detects any abnormal presence of alcohol in the milk, which may indicate contamination or spoilage. The storage temperature of raw milk is also essential to prevent bacterial growth and maintain its quality and stability. High acidity can result from the presence of bacteria, indicating contamination or spoilage. Fat content plays a key role in milk stability, as fats help to form a stable emulsion. Similarly, proteins are important in maintaining the milk's colloidal structure and preventing coagulation or phase separation. In summary, assessing milk stability involves taking into account pH, alcohol, storage temperature, acidity, fat content and protein, to ensure its quality and durability.

The pH of normal fresh cow's milk is around 6.7. This value is largely due to the basic ionizable and dissociable acid groups in proteins, the phosphoric ester groups in caseins, and phosphoric and citric acids.

Take a sampling of the milk and bring it to the boil. If the milk turns sour (forming lumps), the processor must refuse to take this milk, as it will turn sour during pasteurization, and will therefore be unable to withstand the temperatures required to eliminate germs.

Take a 10 centilitre milk sample and mix with 10 centilitres of 60° alcohol. Mix together. If lumps form ("coagulation"), the milk should be rejected, as this indicates the probable presence of germs. However, it is possible that milk that coagulates in the alcohol test can withstand pasteurization. It is therefore not a completely reliable test.

Measuring acidity provides an indication of whether acidification reactions have begun (indicator of lactic acid–fermentation bacteria activity). The advantage of this test is that it's easy to use, inexpensive and gives immediate results. On leaving the udder, healthy milk has a natural acidity of between 15° and 21° Dornic (14° to 16° for goat's milk). On arrival at the dairy, milk acidity is measured to check that fermentation has not begun.

b. **Measuring milk fat content**

The principle of milk fat (MF) separation is based on centrifugation in a butyrometer, where sulfuric acid attacks the milk components, with the exception of the MF. The addition of iso-amyl alcohol promotes fat separation (Figure 9.2).

The procedure for this method is as follows:

- Add 10 ml sulfuric acid ($d = 0.818$) to the butyrometer.
- Using a pipette, add 11 ml of milk to the butyrometer, placing the pipette tip in contact with the butyrometer to avoid premature mixing of milk and sulfuric acid.
- Next, add 1 ml iso-amyl alcohol without mixing the liquids.
- Close the butyrometer tightly and shake vigorously until the mixture is homogeneous.
- At this point, the butyrometer reaches a temperature of around 80°C, due to the mixing of the sulfuric acid with the milk.
- After stirring, centrifuge at 65°C, at 1200 rpm, for 5 minutes.

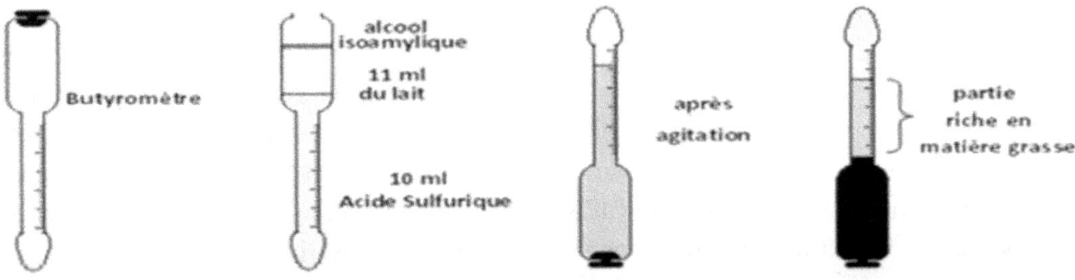

FIGURE 9.2 The GERBER method.

To read the results:

- Remove the butyrometer from the centrifuge and carefully adjust the neck cap to position the lower end of the fat column opposite a numbered mark, moving the column as little as possible.
- The reading should be taken quickly by moving the butyrometer in front of the eyes and noting the lowest level of the meniscus.
- The normal fat content of milk is 40 g/l.

This method quantifies the fat content of milk by centrifugation and reading the fat column level in the butyrometer. The reading is taken quickly to obtain an accurate measurement.

c. **Determination of dry matter content**
The principle of the method is to dry the sample in a crucible at a constant temperature of 103°C in a ventilated oven, at atmospheric pressure, until its weight reaches a practically constant value.

This method works as follows:

- Prepare crucibles by placing in oven at 103 ± 1°C for half an hour, cooling and weighing.
- Weigh 1 ml of sample into the prepared crucibles.
- Dry samples in the oven at 103 ± 1°C for 24 hours, then cool in a desiccator for 30 minutes.

The results can be read according to the following relationship: Percentage of dry matter (MS%) = $(m_1 - m_0) \times 100/m_2$ Where: MS: percentage of dry matter in % m_0: mass of empty crucible (g) m_1: mass of crucible and milk residue (g) m_2: weight of sample taken (g)

This method determines the dry matter content of the sample by calculating the difference in mass between the crucible and the milk residue after drying. Dry matter content is expressed as a percentage.

d. **Microbiological analysis**

The aim of microbiological analyses in milk is to detect the presence of microorganisms, assess their bacteriological quality and verify the safety level of milk, both for raw and pasteurized milk.

For these microbiological analyses, the focus is on a specific parameter: total aerobic mesophilic flora (TAMF). FMAT is a health indicator that measures the number of Colony Forming Units (CFU) present in a product or on a surface.

CFU counts are carried out at a temperature of 30°C, enabling us to distinguish three types of flora: thermophilic flora, which prefer an optimum growth temperature of 45°C; mesophilic flora, which prefer an optimum growth temperature between 20°C and 40°C; and psychrophilic flora, which prefer an optimum growth temperature of 20°C.

Since the culture medium used is favorable to most micro-organisms, with the exception of demanding micro-organisms and strictly anaerobic micro-organisms, it is more appropriate to speak of Mesophilic Aerobic Flora at 30°C rather than "total flora".

The unit of measurement used to quantify micro-organisms is the CFU (Colony Forming Unit), since a colony observable in the agar medium may originate from a single micro-organism, a spore or a combination of micro-organisms.

The analysis procedure is as follows:

- Transfer to a single sterile Petri dish.
- Using a sterile pipette, transfer 1 ml of the sample to be tested (or prepare a dilution series if necessary).
- Pour approximately 15 ml of Plate Culture Agar (PCA) culture medium into each Petri dish.
- Mix the inoculum thoroughly with the culture medium.
- Allow the mixture to solidify.
- Incubate plates for 3 days at 30°C ± 1°C.
- Count and select colonies for confirmation.

These steps make it possible to assess the quantity of micro-organisms present in the milk sample, providing information on its bacteriological quality and safety. Raw cow's milk undergoes a series of microbiological analyses to control its hygienic quality [11]. These analyses are summarized in Table 9.1.

The results of this analysis are shown in the table above, in accordance with the established standard. The analysis is carried out after 72 hours' incubation at a temperature of around 30°C, using PCA.

TABLE 9.1 Raw Milk Analysis

Desired Germ	Media Used	Incubation Temperature	Incubation Time
Mesophilic total	PCA	37°C	24 hours
Total coliforms	VRBL	37°C	24 to 48 hours
Faecal coliform	VRBL	44°C	24 to 48 hours
Yeast and mold	OGA	25°C	3 to 5 days
Staphylococcus aureus	BP	37°C	24 hours
Clostridium SR	VF	37°C	24 à 48 hours

9.3.1.1 *Receiving Milk Method*

Milk reception takes place in several stages. Before the tanks are emptied, a sample is taken for physicochemical and microbiological analyses in the laboratory, as well as rapid tests for acidity, fat content, density and pH.

The acceptance stages are as follows:

- Degassing: The milk passes through a degasser to eliminate odors and gas bubbles present in the milk.
- Filtration: The milk passes through two cylindrical metal filters with pores of increasing diameter to retain unwanted suspended matter such as shavings, plumules, etc. It is important to clean these filters regularly to maintain their efficiency.
- Cooling: Before being stored, milk is cooled to 4°C by passing through plate heat exchangers, where it is cooled using chilled water in a counter-current. The purpose of this cooling is to stop microbial activity.
- Storage: Depending on the state of the milk and the company's activities, milk may be stored or directly processed.

Thermization or pre-pasteurization: This is a less intense heat treatment applied to raw milk for around 15 seconds at temperatures between 83 and 88°C. The aim of this treatment is to destroy the milk's pathogenic bacteria and a large proportion of its common bacterial flora, albeit less intensively than during pasteurization. These various stages guarantee the quality and safety of the milk prior to storage or further processing.

9.4 RESULTS AND DISCUSSION

9.4.1 Physical and Chemical Analysis of Milk

The three Figures 9.3, 9.4 and 9.5 present the results of the analyses for four samples at the three positions determined, Raw milk, Thermized milk; Thermized milk after one hour, such as AC, MG, TP, EST, T, TA.

The physicochemical analysis of raw milk shows that SAFILIAT's raw milk meets standards for all the criteria analyzed. All four samples remained stable in terms of AC, pH, MG, TP, EST and T. The P2 sample was unstable in terms of TA. During the storage phase (thermization after one hour), the TA analysis value decreases to 76%, despite the fact that it poses a threat to the stability of this P2 variety, but is within the company's quality standards.

9.4.2 Microbiological Analysis of Milk

The test was carried out on four samples of raw milk thermized at 84°C for 15 seconds, with the aim of determining the number of CFU present in the samples of raw milk and thermized milk after one hour. The results obtained are shown in the following table:

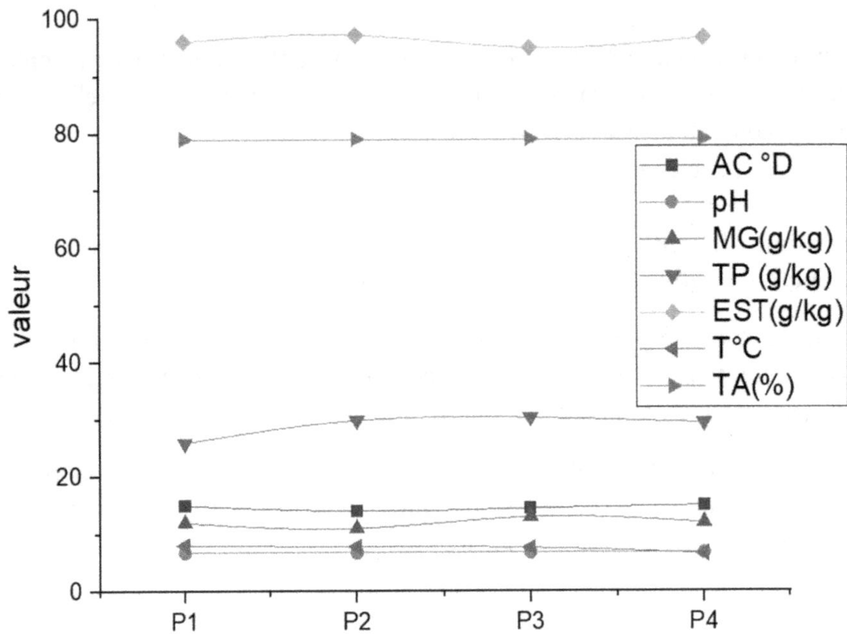

FIGURE 9.3 Results of physicochemical analysis of raw milk.

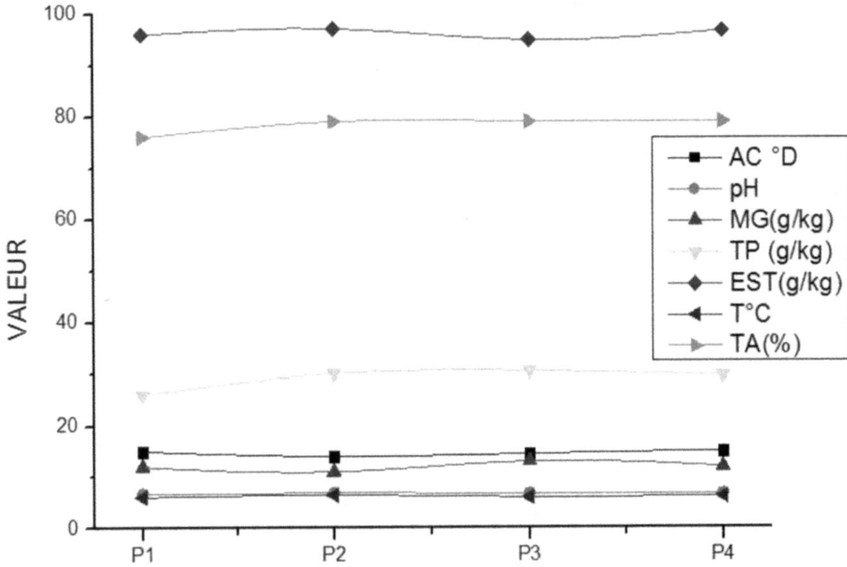

FIGURE 9.4 Results of physicochemical analysis of thermized milk after one hour.

The products analyzed are of satisfactory bacteriological quality (the identification threshold is 3 × 104), and are therefore declared fit for consumption according to company standards. The results also show that thermization was effective in reducing FMAT, with 99.99% reduction in the original LES ETABLES sample, 99.98% reduction in 2 P2 and P3 samples and even a 99.97% reduction in the P4 sample.

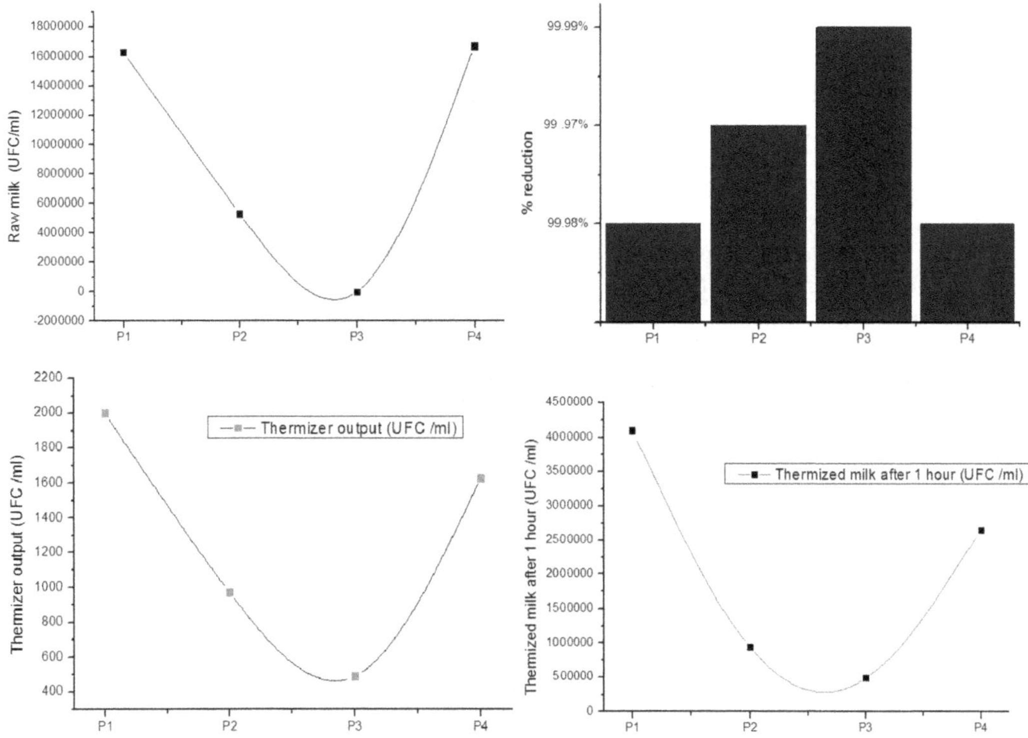

FIGURE 9.5 Number of CFUs present in samples of raw milk and heat-treated milk after one hour.

9.5 CONCLUSION

Thermal processing is a short-term, temporary conservation method that can be followed by pasteurization. Its purpose is to destroy undesirable micro-organisms while preparing the substrate for the growth of beneficial micro-organisms. The values analysed for all samples comply with the company's standards, with the exception of the P2 sample which slightly exceeds these standards, despite similar storage conditions and treatments. To guarantee the stability of raw milk, several factors need to be taken into consideration. Internal factors include storage conditions, which play a crucial role. Inappropriate storage temperatures, particularly warm ones, can encourage the growth of alcohol-producing micro-organisms, leading to an increase in the amount of alcohol in the milk. Furthermore, raw milk naturally contains lactic acid bacteria, which can cause fermentation. If raw milk is not stored correctly, this can lead to contamination and an increase in the alcohol content of the milk.

External factors, such as external contamination, can also influence the alcohol content of raw milk. If milk is exposed to environments containing alcohol-producing yeasts or bacteria, external contamination can occur, leading to an increase in alcohol levels. This can occur as a result of poor milk storage or contact with contaminated surfaces. In addition, the lactation stage of cows can also have an impact on alcohol levels in raw milk, although this can vary from cow to cow. Assessing the stability of raw milk on receipt is essential to guarantee food safety, dairy product quality and compliance with regulatory standards. It enables corrective action to be taken where necessary to improve raw milk production and handling practices.

REFERENCES

[1] Kirat, (2007). *Les conditions d'émergence d'un système d'élevage spécialisé enengraissement et ses conséquences sur la redynamisation de l'exploitation agricole et la filière des viandes rouges bovines - Cas de la Wilaya de Jijel en Algérie.* Montpellier (France): CIHEAM-IAMM.

[2] Faye, B., et Loiseau, G., (2002). *Gestion de la sécurité des aliments dans les pays en développement actes de l'atelier international CIRAD on sources de contamination dans les filières laitières et exemples de démarches qualité*, Montpellier: 11–13.

[3] Mazoyer, (2002). Larousse agricole, le monde agricole au XXIème siècle. Mathilde: 767.

[4] Arie, F., Sri, Kumalaningsh et Ariesta, W., (2012). Process engineering of drying milk powder with foam mat drying method. *Journal of Basic and Applied Scientific Research* 2(4), 3588–3592.

[5] Bruneliere, R., et Zabi, A. *Reconstruction of the signal amplitude of the CMS electromagnetic calorimeter.* CERN-CMS-NOTE-2006-037, 2006.

[6] Chatellier, V. (2019, June). La planète laitière et la place de l'Afrique de l'Ouest dans la consommation, la production et les échanges de produits laitiers. In 3. Rencontres internationales:" Le lait, vecteur de développement" (pp. 32–p).

[7] Jeantet, R. C. T., Mahaut, M, Schuck P, Brulé G., (2008). *Les produits laitiers (Technique et documentation).* Lavoisier (Ed.), Paris. ed.

[8] Philippe Dudez, Cécile Broutin (16/01/2003). Techniques/Technologies Mise en œuvre, GRET(France), Global; Mots clés LAIT; CONTROLE DE QUALITE; PRODUIT LAITIER; QUALITE.

[9] Organisation internationale de normalisation (ISO), (2003). Norme ISO 4833 (F). Microbiologie des aliments. Méthode horizontale pour le dénombrement des micro-organismes. Technique par comptage des colonies à 30°C. ISO, Paris, 9pp.

[10] Davies, D. T., White, J. C. D., 1958. The relation between the chemical composition of milk and the stability of the caseinate complex. II. Coagulation by ethanol. *Journal of Dairy Research* 25, 256–266.

[11] https://www.rapport-gratuit.com/reception-refroidissement-et-stockage-du-lait-cru/.

CHAPTER 10

Characterization of Climatic Variables Inside a Smart Mixed Solar Dryer

Mohammed Benchrifa

University Mohammed V in Rabat, Morocco
Ibn Tofaïl University—Kenitra-University Campus, Kenitra, Morocco

Jamal Mabrouki and Rachid Tadili

University Mohammed V in Rabat, Morocco

10.1 INTRODUCTION

Drying fruits and vegetables is the oldest methods of preserving agricultural products, because the quality of dried produce is significantly improved. Drying agricultural produce is primarily aimed at reducing the moisture content to a level that can be stored safely for a prolonged period [1–6]. It also leads to a reduction in weight as well as volume, which reduces packaging, storage and transport costs [7–12]. The choice of foodstuffs to be dried was based on the tomato. Firstly, because of its widespread cultivation throughout the world, for its richness in nutrients. Secondly, because of their irregular availability. However, their preservation through drying allows continuity in their availability. Solar drying is considered to be the most widely used solar energy system [13]. Drying fruit, vegetables and meat is one of the most energy-intensive processes in the food industry and is a better way of reducing post-harvest costs and losses [14]. For centuries, solar drying has been practised in the open air all over the world. The process was used to dry grain, fruit, meat, fish and other food products intended for consumption [15–17]. However, the traditional drying method suffers from many problems, including the impossibility of controlling the drying process correctly, time uncertainty, expensive labour costs, the large surfaces required, and the risk of being infected by insects [18]. Thus, the objective of our work is to realize and experiment a mixed solar dryer, to make it intelligent by adding an embedded system and then to evaluate the different climatic variables inside the drying chamber and, finally, to study the impact of this system on the drying of the tomatoes.

DOI: 10.1201/9781003436218-10

10.2 MATERIALS AND METHODS

10.2.1 Embedded Systems

An embedded system is a complex system that integrates software and hardware designed together to provide specific functionality. It usually contains one or more microprocessors to run a set of programs defined at conception and stored in memories. The hardware system and the application (software) are intimately linked and immersed in the hardware and are not as easily discernible as in a conventional desktop PC (Personal Computer) environment [19].

An embedded system is self-contained and does not have standard inputs/outputs such as a keyboard or computer monitor. As opposed to a PC, the Human Machine Interface (HMI) of an embedded system can be as simple as a flashing Light Emitter Diode (LED) or as complex as a real-time night vision system; Liquid Crystal Display (LCD) displays of generally simple structure are commonly used.

Embedded systems generally operate in real time (RT): computational operations are then performed in response to an external event (hardware interrupt) [20].

In our case we used an Arduino 'mega' (Figure 10.1) board equipped with a LM35 temperature sensor to measure the internal temperature of the dryer and a relay that turns the fan on or off according to the instructions given. However, when the temperature exceeds 40°C, the fan is automatically activated to circulate thermal energy throughout the dryer, but when the temperature falls below 40°C the fan automatically shuts down, knowing that the test is repeated after every minute (Figure 10.2).

FIGURE 10.1 Used embedded system.

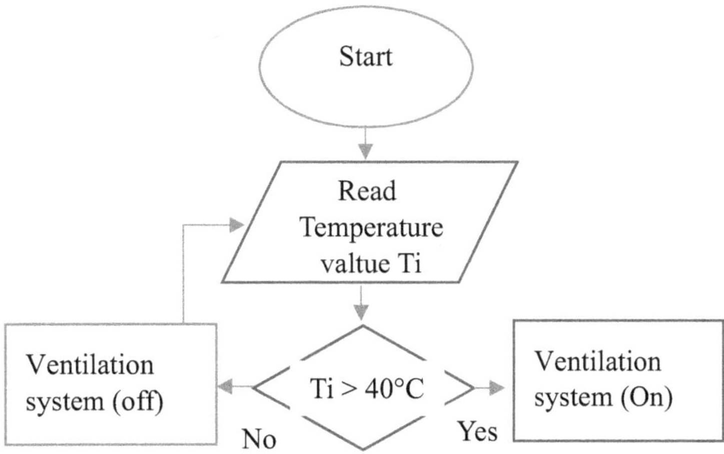

FIGURE 10.2 Flowchart of follow-up steps.

10.2.2 Characteristics of the Tested Solar Dryer

Drying is a way of conserving agricultural produce [21–23]. Drying using fossil fuels consumes a lot of energy. As the price of oil has been rising in recent years and is expected to continue to rise, it is important to develop an energy-free drying system: solar drying is one such system. It uses only the energy of the Sun, which is free and inexhaustible.

Solar dryers are generally classified into several categories, depending on the heating method or the way they operate. Thus, there are natural dryers, direct solar dryers, indirect solar dryers, mixed solar dryers and hybrid solar dryers [24–27].

By definition, natural dryers use the Sun and air directly, whose action is neither particularly favoured nor controlled. The product is spread out on racks or placed on the ground. These dryers are very cheap, but require regular and sustained human intervention: protection or collection of the product in case of rain, frequent mixing to avoid overheating of the upper layer and to allow the lower layer to dry. This type of dryer is often traditional in farming communities, to address the problem of temporary storage of the product while awaiting sale or consumption [27–32].

In direct solar dryers, sunlight dries the food directly. This type of dryer requires, for example, a wooden or cardboard box with holes at the bottom and top to allow cold air to enter from the bottom and hot air to leave from the top. This box contains the racks where the food will be dried and is covered with glass to increase the greenhouse effect [33–35].

Mixed solar dryers are a combination of direct and indirect dryers. In these dryers, the heat required for drying is provided by the action of the Sun's rays, which strike the products in the direct section and heat the air. The heat and water vapor transfers are complex and not well known. The mixed solar dryer consists of two parts: the lower part is the direct dryer, when the Sun heats the internal air of this part, the hot air passes to the drying chamber through holes, that is the indirect dryer part. The air, circulating through the fruit and vegetables, dehydrates them little by little and comes out of the upper part, loaded with water vapor.

We have chosen the mixed solar dryer because of its advantages and mainly because of its characteristic of coupling between direct and indirect solar dryers. This coupling will allow us to benefit from the specificity of each of these dryers at the same time, namely the conservation of the qualities of the dried agricultural product such as its color, flavor, nutritional value (in particular, vitamins A and C), as well as reaching high temperatures to dry products that have a high-water content such as wood. [7]

The direct part of the dryer that is made is the double glazing, which is a glass wall made up of two panes of glass separated by a layer of still air, known as the "air gap". The advantage of double glazing is that it provides better thermal insulation, as the air space is a good insulator, much better than the glass itself. Double glazing thus reduces condensation and heat loss.

The thicknesses are often referred to as a/b/c. Where a, b and c are the thicknesses in millimeters of the elements (outer pane, air space, inner pane). Common double glazing is 4/3/4/ [7]. In the base of the direct part an insulation (polystyrene) and above an absorber to convert solar radiation into heat have been placed.

The indirect part consists of the drying chamber, which is a cabinet in the form of a parallelepiped. Its volume is (120 cm × 60 cm × 60 cm) or 432 × 103 cm^3. Its walls are made of TRESPA wood with a thickness of 0.8 cm. TRESPA is an extremely versatile material. The panels can easily be combined with other materials. Manufactured under high pressure and at high temperature from a mixture of up to 70% wood or cellulose fibers and thermosetting resins, TRESPA is an extremely stable, dense board. The dryer has a glass door with an aluminum frame to put product into the drying chamber and to recover it at the end of the operation [7]. Three rectangular trays are placed on the drying chamber. The mesh rack is surrounded by wood and has a size of (52 cm × 61 cm). Its openings are (2.5 mm × 2.5 mm). The distance between the trays is 32 cm. This space is large enough to allow for better air circulation (Figure 10.3).

10.2.3 Steps Followed for the Collection of Measurements

In order to carry out the different tests we used several experimental devices with different measuring instruments. A Kepp and Zonen pyranometer was used to measure the global solar radiation on a horizontal plane, and four thermocouples were used to measure the temperature at different parts of the dryer. These thermocouples convert temperature into voltage. When two electrical wires made of different metals are joined, a very low electrical voltage is generated. This voltage is a function of the temperature of the junction and the type of metals that make up the thermocouple wires. Since the temperature characteristics of metals are well known, it is easy to convert the voltage produced by the junction to its temperature.

10.3 RESULTS AND DISCUSSION

The drying of medicinal and aromatic plants often remains a poorly controlled and costly process for producers. The water content of a medicinal plant must be reduced rapidly to

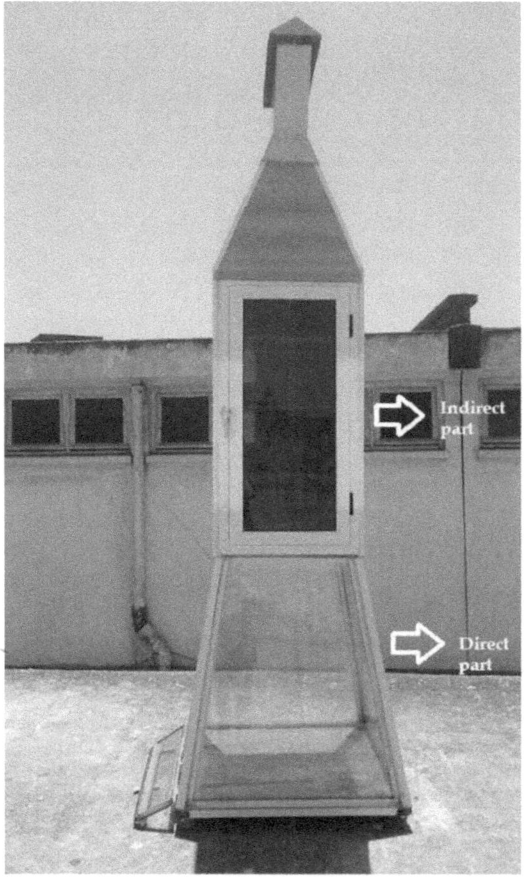

FIGURE 10.3 Experimented solar dryer.

preserve it, but at a low temperature so as not to deteriorate its active principles. In the food industry in general, the optimization of the drying operation must meet two essential requirements, which are the limited consumption of the necessary energy and the preservation of the quality of the dried product. In this part we are particularly interested in the experimental study of a mixed solar dryer equipped with an intelligent system that controls the ventilation.

The solar drying prototype ensures the full use of thermal energy from the Sun. In order to study the usefulness of this system, it is necessary to study the different variables inside the system. Thus, with the help of several instruments, the radiation inside the direct part of the dryer, the temperature of the glazing and the temperature of the absorber as well as the temperature of the area on the first and third floor of the indirect part were measured. The results of these measurements are shown in Figure 10.4.

According to the figure, which represents the variation of the global irradiation on horizontal plane inside the direct part of the dryer, we notice a considerable decrease of the radiation since it decreases from 528 W/m^2 outside to 305 W/m^2 inside which represents more than 40% of lost solar radiation. This loss is mainly due to the inclination of the glazing, which is greater than 70°, and to the nature of the glazing, which is double-glazed.

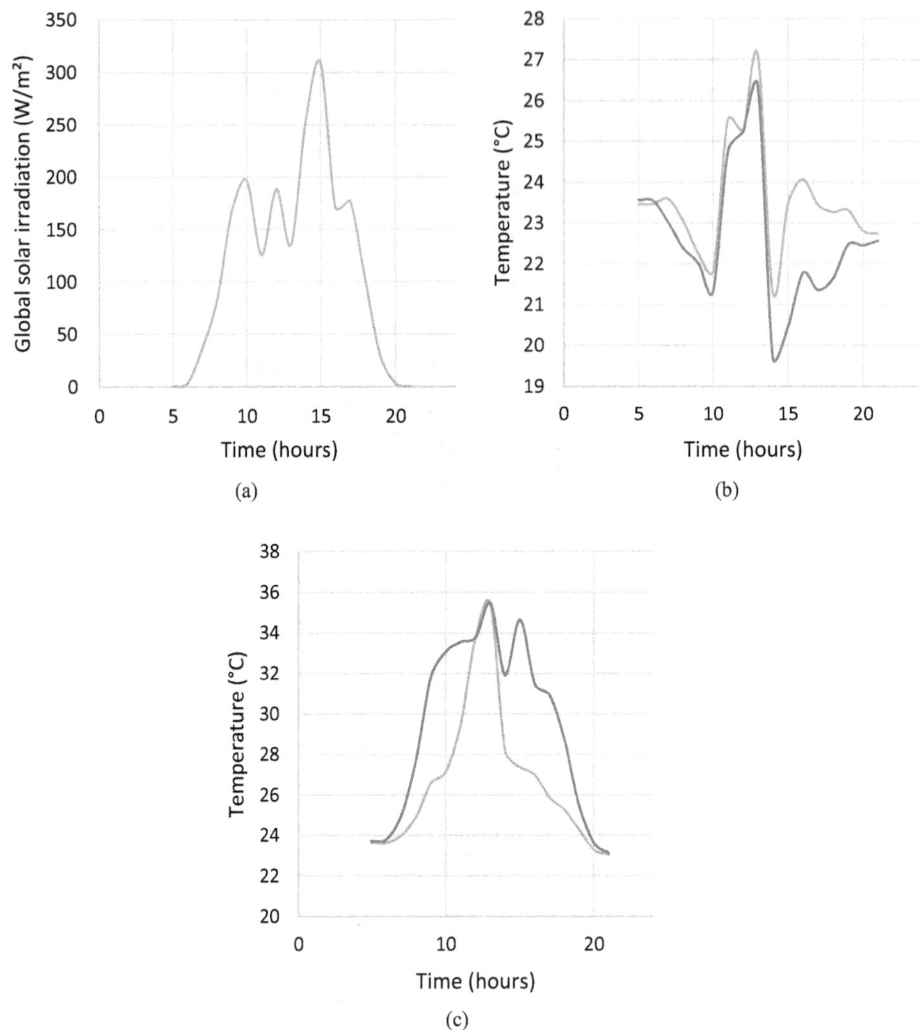

FIGURE 10.4 Solar radiation and temperature variation inside the solar dryer as function of time. (a) Solar radiation hourly variation; (b) Variation of temperature inside the indirect part; (c) Variation of temperature inside the direct part.

When it comes to temperature variation, it can be seen that in the direct part the temperature of the absorber and the internal glazing can reach 35°C, while in the indirect part the temperature decreases from one floor to another. This decrease is mainly due to the loss of thermal energy through the uninsulated side walls.

10.4 CONCLUSION

The energy we receive from the Sun on Earth is the energy the Sun emits primarily in the form of heat and light. Since the Sun is an inexhaustible source of energy and its daily utilization is essential, it is considered the main energy resource that ensures the survival of life on Earth. The main use of this energy is to capture sunlight and convert it into heat or electricity. Sun drying varies from region to region, and there are different ways to ensure the success of this process. The designated area should have high ambient temperature and high solar radiation intensity. Morocco receives plenty of sun on most days of the year, so drying time and product quality retention are his two basic criteria for choosing the right drying method. Drying is one of the basic processes that effectively increases the shelf life of products in the food industry; in recent years, the use of solar dryers has seen great development. Thus, the aim of our work was to study the variation of temperature and radiation inside a mixed solar dryer.

The experimental study showed a significant reduction in the radiation passing through the glass, from 528 W/m^2 outside to 305 W/m^2 inside, which represents more than 40% of the lost solar radiation. This loss is mainly due to the inclination of the glazing, which is greater than 70°, and to the nature of the glazing, which is double-glazed. In terms of temperature variation, in the direct part, the temperature of the absorber and the internal glazing can reach 35°C, while in the indirect part, the temperature decreases from floor to floor. This decrease is mainly due to the loss of thermal energy through the uninsulated side walls.

These results show that there are several errors in the design of this solar dryer. In order to improve the efficiency of the solar dryer, it is necessary to make several improvements to the system, such as insulating the drying chamber and reducing the inclination of the glazing in the direct part of the dryer.

These results show that there are several flaws in the design of this solar dryer. In order to improve the efficiency of the solar dryer, it is necessary to make several improvements to the system, such as insulating the drying chamber and reducing the inclination of the glazing in the direct part of the dryer.

REFERENCES

1. Hii, C. L., Law, C. L., & Cloke, M. (2008). Modelling of thin layer drying kinetics of cocoa beans during artificial and natural drying. *Journal of Engineering Science and Technology*, *3*(1), 1–10.
2. Röser, D., Mola-Yudego, B., Sikanen, L., Prinz, R., Gritten, D., Emer, B., … & Erkkilä, A. (2011). Natural drying treatments during seasonal storage of wood for bioenergy in different European locations. *Biomass and Bioenergy*, *35*(10), 4238–4247.
3. González, C. M., Gil, R., Moraga, G., & Salvador, A. (2021). Natural drying of astringent and non-astringent persimmon "Rojo Brillante". Drying kinetics and physico-chemical properties. *Foods*, *10*(3), 647.
4. Sontakke, M. S., & Salve, S. P. (2015). Solar drying technologies: A review. *International Refereed Journal of Engineering and Science*, *4*(4), 29–35.
5. Gigler, J. K., van Loon, W. K., Van den Berg, J. V., Sonneveld, C., & Meerdink, G. (2000). Natural wind drying of willow stems. *Biomass and Bioenergy*, *19*(3), 153–163.
6. Tiris, C., Ozbalta, N., Tiris, M., & Dincer, I. (1994). Performance of a solar dryer. *Energy*, *19*(9), 993–997.
7. Nukulwar, M. R., & Tungikar, V. B. (2021). A review on performance evaluation of solar dryer and its material for drying agricultural products. *Materials Today: Proceedings*, *46*, 345–349.
8. Kabeel, A. E., & Abdelgaied, M. (2016). Performance of novel solar dryer. *Process Safety and Environmental Protection*, *102*, 183–189.

9. Hegde, V. N., Hosur, V. S., Rathod, S. K., Harsoor, P. A., & Narayana, K. B. (2015). Design, fabrication and performance evaluation of solar dryer for banana. *Energy, Sustainability and Society*, *5*(1), 1–12.
10. Fudholi, A., Sopian, K., Yazdi, M. H., Ruslan, M. H., Gabbasa, M., & Kazem, H. A. (2014). Performance analysis of solar drying system for red chili. *Solar Energy*, *99*, 47–54.
11. Janjai, S., & Tung, P. (2005). Performance of a solar dryer using hot air from roof-integrated solar collectors for drying herbs and spices. *Renewable Energy*, *30*(14), 2085–2095.
12. Rezaei, M., Sefid, M., Almutairi, K., Mostafaeipour, A., Ao, H. X., Dehshiri, S. J. H., … & Techato, K. (2022). Investigating performance of a new design of forced convection solar dryer. *Sustainable Energy Technologies and Assessments*, *50*, 101863.
13. Moussaoui, H., Bahammou, Y., Idlimam, A., Lamharrar, A., & Abdenouri, N. (2019). Investigation of hygroscopic equilibrium and modeling sorption isotherms of the argan products: A comparative study of leaves, pulps, and fruits. *Food and Bioproducts Processing*, *114*, 12–22.
14. Zhang, Z., Kim, C. S., & Kleinstreuer, C. (2006). Water vapor transport and its effects on the deposition of hygroscopic droplets in a human upper airway model. *Aerosol Science and Technology*, *40*(1), 1–16.
15. Martin, E. C. (1958). Some aspects of hygroscopic properties and fermentation of honey. *Bee world*, *39*(7), 165–178.
16. Bennamoun, L., & Belhamri, A. (2003). Design and simulation of a solar dryer for agriculture products. *Journal of Food Engineering*, *59*(2–3), 259–266.
17. Amer, B. M. A., Hossain, M. A., & Gottschalk, K. (2010). Design and performance evaluation of a new hybrid solar dryer for banana. *Energy Conversion and Management*, *51*(4), 813–820.
18. Ekechukwu, O. V., & Norton, B. (1999). Review of solar-energy drying systems II: An overview of solar drying technology. *Energy Conversion and Management*, *40*(6), 615–655.
19. Ayensu, A. (1997). Dehydration of food crops using a solar dryer with convective heat flow. *Solar Energy*, *59*(4–6), 121–126.
20. Bolaji, B. O., & Olalusi, A. P. (2008). Performance evaluation of a mixed-mode solar dryer. *AU Journal of Technology*, *11*(4), 225–231.
21. Kamarulzaman, A., Hasanuzzaman, M., & Rahim, N. A. (2021). Global advancement of solar drying technologies and its future prospects: A review. *Solar Energy*, *221*, 559–582.
22. Ahmadi, A., Das, B., Ehyaei, M. A., Esmaeilion, F., Assad, M. E. H., Jamali, D. H., … & Safari, S. (2021). Energy, exergy, and techno-economic performance analyses of solar dryers for agro products: A comprehensive review. *Solar Energy*, *228*, 349–373.
23. Kong, D., Wang, Y., Li, M., Liu, X., Huang, M., & Li, X. (2021). Analysis of drying kinetics, energy and microstructural properties of turnips using a solar drying system. *Solar Energy*, *230*, 721–731.
24. Vigneshkumar, N., Venkatasudhahar, M., Kumar, P. M., Ramesh, A., Subbiah, R., Stalin, P. M. J., … & Kriuthikeswaran, M. (2021). Investigation on indirect solar dryer for drying sliced potatoes using phase change materials (PCM). *Materials Today: Proceedings*, *47*, 5233–5238.
25. El-Mesery, Hany S., El-Seesy, Ahmed I., Hu, Zicheng, et al. (2022). Recent developments in solar drying technology of food and agricultural products: A review. *Renewable and Sustainable Energy Reviews*, *157*, 112070.
26. Philip, N., Duraipandi, S., & Sreekumar, A. (2022). Techno-economic analysis of greenhouse solar dryer for drying agricultural produce. *Renewable Energy*, *199*, 613–627.
27. Natarajan, S. K., Elangovan, E., Elavarasan, R. M., Balaraman, A., & Sundaram, S. (2022). Review on solar dryers for drying fish, fruits, and vegetables. *Environmental Science and Pollution Research*, *29*, 40478–40506.
28. Benchrifa, M., Mabrouki, J., Elouardi, M., Azrour, M., & Tadili, R. (2023). Detailed study of dimensioning and simulating a grid-connected PV power station and analysis of its environmental and economic effect, case study. *Modeling Earth Systems and Environment*, *9*, 53–61.

29. Benchrifa, M., & Mabrouki, J. (2022). Simulation, sizing, economic evaluation and environmental impact assessment of a photovoltaic power plant for the electrification of an establishment. *Advances in Building Energy Research, 16*, 1–18.
30. Essalhi, H., Benchrifa, M., & Tadili, R. (2022). Effect of natural and forced ventilation on drying of Kiwi in an indirect solar dryer. In *Sustainable energy development and innovation* (pp. 177–180). Cham: Springer.
31. Benchrifa, M., Essalhi, H., Tadili, R., & Nfaoui, H. (2022). Estimation of daily direct solar radiation for Rabat. In *Sustainable energy development and innovation* (pp. 629-634). Cham: Springer.
32. Idrissi, A., Benchrifa, M., Tadili, R., Naciri, M., & Madrab, S. E. (2022). Experimental investigation of solar still with corrugated absorber surface for domestic desalination use. In *Sustainable energy development and innovation*, edited by Ali Sayigh, (pp. 187–191). Cham: Springer.
33. Mabrouki, J., Fattah, G., Al-Jadabi, N., Abrouki, Y., Dhiba, D., Azrour, M., & Hajjaji, S. E. (2022). Study, simulation and modulation of solar thermal domestic hot water production systems. *Modeling Earth Systems and Environment, 8*(2), 2853–2862.
34. Mabrouki, J., Azoulay, K., Elfanssi, S., Bouhachlaf, L., Mousli, F., Azrour, M., & Hajjaji, S. E. (2022). Smart system for monitoring and controlling of agricultural production by the IoT. In *IoT and smart devices for sustainable environment* (pp. 103–115). Cham: Springer.
35. Mabrouki, J., Fattah, G., Kherraf, S., Abrouki, Y., Azrour, M., & El Hajjaji, S. (2022). Artificial intelligence system for intelligent monitoring and management of water treatment plants. In *Emerging real-world applications of internet of things*, edited by Anshul Verma, Pradeepika Verma, Yousef Farhaoui, Zhihan Lv, (pp. 69–87). CRC Press.

Index

Pages in *italics* refer to figures and pages in **bold** refer to tables.

U

V

W

For Product Safety Concerns and Information please contact our EU
representative GPSR@taylorandfrancis.com
Taylor & Francis Verlag GmbH, Kaufingerstraße 24, 80331 München, Germany

www.ingramcontent.com/pod-product-compliance
Lightning Source LLC
LaVergne TN
LVHW081320110826
845149LV00006B/1551